The China Environment Yearbook, Volume 2

The Chinese Academy of Social Sciences Yearbooks: Environment

Volume 2

BEIJING
2008

The China Environment Yearbook, Volume 2

Changes and Struggles

Edited by

YANG Dongping
Friends of Nature

BRILL

LEIDEN • BOSTON
2008

This yearbook is the result of a co-publication agreement between Social Sciences Academic Press and Koninklijke Brill NV. These articles were translated into English from the original《环境绿皮书2006年：中国环境的转型与博弈》(*Huanjing Lü Pi Shu 2006 Nian: Zhongguo Huanjing de Zhuanxing yu Boyi*) with financial support from China Book International, supported by the General Administration of Press and Publication and the Information Office of the State Council of China.

This book is printed on acid-free paper.

ISSN 1872-7212
ISBN 978 90 04 16800 8

Koninklijke Brill NV incorporates the imprints Brill, Hotei Publishing, IDC Publishers, Martinus Nijhoff Publishers and VSP.

PRINTED IN THE NETHERLANDS

CONTENTS

GENERAL REPORT

PART ONE

THE ENVIRONMENT AND SOCIETY

PART FOUR

FORESTS

PART FIVE

APPENDIX

LIST OF FIGURES AND TABLES

Figures

Tables

PREFACE

The China Environment Yearbook, Volume 2: Changes and Struggles is the English language version of the second volume in the original Chinese *Green Book of Environment* series (*Huanjing Lüpishu*). In this volume, the Green Culture Institute of the International Academy of Chinese Culture, popularly known as Friends of Nature, adheres to the guiding principle laid down in *The China Environment Yearbook (2005): Crisis and Breakthrough of China's Environment*, the first English language volume in the series: to observe and record the past year's major events in environmental protection from an NGO's perspective. By maintaining its distinct tone—more independent than the "official" governmental voice and more personal than a scholarly text—the book helps readers better understand China's environmental changes, problems, challenges, experiences, and lessons during a time when China is moving towards realizing a more sustainable form of development.

We have been encouraged by the praise we received from *The China Environment Yearbook (2005)*, yet we face the challenge of meeting increasingly high expectations in our follow-up. As with last year's book, the contributors to this year's edition are from the front lines of environmental protection work. Notable environmental specialists, scholars, NGO workers, and journalists are represented. Their articles contain serious reflections on environmental problems, distilling long hours of work into relatively compact texts.

In compiling the book you now hold, we have tried to incorporate design innovations to make this volume more reader-friendly. For instance, we have added a chapter of Annual Indexes that supplies an array of recent indexes allowing for quick comparison and analysis. We hope that this helps readers to more quickly and easily see the trends at work that we cover in more detail in the book's chapters.

The completion of this book, and its improvements, is attributed to the precious suggestions of our readers. Taking our readers' feedback into consideration, we have grouped the articles according to topics, made articles more compact, and added tables and figures to provide readers more ways of understanding the material.

We sincerely hope that readers find this volume to be a useful tool in understanding China's current environmental situation. We encourage interested readers, scholars, activists, and policy-makers to contact our organization to learn more about what we do, and to explore avenues of cooperation on the worthy cause of ensuring a sustainable future for China's environment, as well as for society as a whole.

The Green Book Project of Friends of Nature
January 17, 2007

A BRIEF INTRODUCTION TO FRIENDS OF NATURE

Established on March 31, 1994, Friends of Nature (FON) is the first officially registered environmental nongovernmental organization (NGO) in China. The four founding members are Liang Congjie, Yang Dongping, Liang Xiaoyan, and Wang Lixiong. Liang Congjie serves as its president. He also a member of the Chinese People's Political Consultative Conference (CPPCC) and a professor with the Institute of Green Culture of the International Academy of Chinese Culture. The vice president, Yang Dongping, is a professor of education from Beijing Institute of Technology (BIT), which sets as its mission the diffusion of green culture and promotion of sustainable development.

As an NGO, Friends of Nature complies with Chinese laws and plays a role in supervising the implementation of governmental policies. Over the past decade more than eight thousand people have joined FON; currently FON has three thousand active members and thirty university student-body groups. FON has won fifteen national and international awards, including the Asian Environmental Award, the Earth Award, the Giant Panda Award, and the Ramon Magsaysay Award (one of the leading humanitarian awards granted in Asia). Chapters of FON have been established in Nanjing, Guangdong, Shenzhen, and other major cities. More than ten new environmental NGOs have been set up by FON members. Friends of Nature has made great contributions to environmental conservation and has become an influential NGO in China.

For more information, please visit our website at www.fon.org.cn.

Be a Volunteer for Friends of Nature

Mere criticism and complaints cannot solve all problems. To make lasting change, we must take action. If you want to be a volunteer involved in activities organized by Friends of Nature, please write to volunteers@fon.org.cn.

Become a Member of Friends of Nature

Friends of Nature needs your help and participation. Please join us. The more members we have, the more environmental work we can do. Please contact us at membership@fon.org.cn.

Donate to Friends of Nature

Our various programs for environmental protection need financial support. Your donations to Friends of Nature will be highly appreciated. If you would like to make a donation, please contact office@fon.org.cn.

Contact Friends of Nature

Address: No. 53 Ganyu Street, Room 368
East District
Beijing, China
100009
Tel: 0086–10–6523 2040/6512 0929/6512 0827/6512/0937
Fax: 0086–10–6528 6069
Email: office@fon.org.cn
Website: www.fon.org.cn

ACKNOWLEDGMENTS

The English edition of *The China Environment Yearbook, Volume 2: Changes and Struggles* was produced as a result of arduous labor from dozens of highly committed and talented volunteers and scholars. Our deepest admiration and praise are reserved for the article contributors from Friends of Nature who put together yet another monumental work on environmental issues in the original Chinese *Green Book of Environment*. Additionally, gratitude is owed to an indispensible group of researchers who offered informational support used in the text: Liu Yanyan, Zhu Jinshan, Huang Chaoni, Wu Tianyu, Zu Shuxian, Meng Zhibin, Zhang Kebin, Lin Jingxing, Chen Honghan, Xing Liting, Li Ang, Liu Xiao'e, Liu Zijun, Zhang Boju, and volunteers from the Environmental Planning Association of Beijing Forestry University. Holding a shared dream and goal with Friends of Nature, they gave selflessly in supporting this book project.

The task of translation could not have been accomplished without the hard work of dozens of volunteers hailing from the School of English and International Studies of Beijing Foreign Studies University (BFSU). The following translators made the transition from the Chinese *Green Book* to *The China Environment Yearbook, Volume 2* possible: Ai Di, Cai Jindong, Cai Jing, Chen Chen, Chen Ji, Chen Jin, Fu Qi, Geng Jiamin, Guo Ying, Hu Yi, Ji Fang, Kang Qianqian, Kong Qinghua, Li Qi, Liao Jin, Lin Yun, Liu Hao, Ma Shuangshuang, Meng Qinghai, Mo Jiao, Mu Xiaomei, Shen Gaochao, Song Rui, Song Song, Sun Jingyi, Wan Ye, Wang Jing, Wang Xiuqiong, Wen Yi, Xia Meifang, Yang Yuan, Yi Yangge, Zhai Hongxia, Zhang Liqun, Zhang Qi, Zhang Xin, Zhang Weiping, Zhao Yinghui, Zheng Yibing, and Zhu Jinshan.

The first-round translation was revised by Helen Wylie Bartle, W. Chad Futrell, Sara Grimes, Elizabeth Schultz, Dai Ning, Deng Xiaowen, Dou Wei, Fu Meirong, Jia Ning, Ke Ke'er, Liang Hong, Shen Changying, Song Ying, Wang Lili, Wang Wenli, Yi Rui, Zhai Zheng, and Zhang Baiqing.

Titus Levi conducted the final editing for the whole book with a great deal of help from Professor Wang Lili of BFSU. Dr. Levi edited every chapter carefully, improving them not only in language but also organizational structure. He also provided "translator's notes" and examples that help to explain and clarify many points on Chinese experience and methods that may not be familiar to Western readers. The index was prepared by Ma Tianjie, an environmental activist who is now a graduate student at American University. We are indebted to Mr. Ma for his hard work in adding this important tool to the volume.

We are honored that Judith Shapiro and Guobin Yang have agreed to give their time and talents serving on the international advisory board for the English edition of the *Green Book*. We look forward to working with them on future volumes to make the work of environmental scholars, NGO members, and volunteers more accessible to an international English-reading audience.

Finally, we would like to thank all individuals and organizations who have given their support to the Green Book translation project. We hope that you will continue to give your attention to the project and become more involved in environmental protection, a common cause for all citizens who cherish nature and have a strong sense of social responsibility.

The Green Book Project of Friends of Nature
Social Sciences Academic Press, Beijing
Brill, Leiden

LIST OF CONTRIBUTORS

Throughout this volume, Chinese names are ordered according to standard practice in China, with surnames preceding given names. To clarify this ordering, surnames are in all capital letters in the following list of contributors.

CHENG Yun is a graduate student from Beijing Forestry University. She worked as a coordinator for the Xinjiang Conservation Fund (XCF).

FAN Xiao is a general engineer with the Geological Exploitation Bureau of Sichuan Province's geological exploitation team. He is also professor and vice dean at Sichuan Tourism Normal University.

FENG Yongfeng is a reporter for *Guangming Daily*. As a member of Friends of Nature, he has long been concerned about environmental protection in China and has published numerous articles on this subject.

W. Chad FUTRELL is a PhD candidate at Cornell University. Currently he is working in Beijing as a volunteer for Friends of Nature while writing his doctoral dissertation on international cooperation in environmental protection. Futrell studied Korean and the environmental movement in South Korea and participated in the Inter-University Exchange Program between Cornell and Tsinghua University.

GUO Xiaojun is a reporter from *The Beijing News* and has long given attention to China's environmental problems.

HU Kanping served as director of the editorial board of *China Green Times*, and editor-in-chief of *Chinese Forestry* magazine. Currently he is deputy editor-in-chief of *Environmental Protection* magazine, published by the State Environmental Protection Administration of China (SEPA). He is also a council member of the Xinjiang Conservation Fund (XCF). As one of Friends of Nature's senior members, Hu was one of the first to show concern for environmental protection in China.

HUO Daishan is founder and president of the Huaihe River Guardians, an NGO set up for the protection of the Huaihe River, the major west-to-east river system running north of the Yangtze River. Before founding his NGO, Huo worked as a photographer for *Zhoukou Daily*. In 1998 he quit this job and began to give all his attention to investigations of pollution in the Huaihe River Basin. He publicized the pollution problems that continue to plague the Huaihe, even after ten years of government intervention intended to control and reduce pollution in the river basin. His reporting on "cancer villages" has been among the most revealing and shocking environmental reporting in recent years.

LI Xi is associate professor from the School of Politics and Public Administration of Wuhan University. Her research focuses on water and land resource management.

LIU Mei, an instructor from Wuhan University, focuses her research on the green economy and sustainable agricultural development.

LIU Shurun is a specialist in animal husbandry in Inner Mongolia and Tibet. He has done research and teaching at the Grass Institute of the Chinese Academy of Agriculture, Grass Station of Inner Mongolia, Inner Mongolia University, and Inner Mongolian Educational College. His fields cover ecology, cash crops, and eco-tourism. He has also been engaged in evaluating and setting up nature preserves in Inner Mongolia, Xinjiang, Tibet and Northeast China.

MA Jun is director of the Institute of Public and Environmental Affairs (IPE). He worked as a reporter before founding IPE. His published works include *Water Crisis in China*. Ever since being listed as one of the 100 most influential persons in the world by *Time* magazine in May, 2006, Ma has been renowned for his work in protecting water supplies in China. Recently he captured headlines by launching an unprecedented online survey of water pollution, which listed 2,500 seriously polluting enterprises.

MA Zhong is a professor at Renmin University of China and vice dean of the School of Environmental and Natural Resources at the university. As a well-known environmental economics expert, Ma serves on SEPA's Science and Technology Consultant Committee, and is a member of the Environmental Economics Working Group with the

China Council for International Cooperation on Environment and Development (CCICED). He is also editor of *China Environmental Science* magazine.

NING Chen is a graduate student from the Chinese University of Political Science and Law and a member of the Center for Legal Assistance to Pollution Victims in China.

JUDITH SHAPIRO is the director of Global Environmental Politics programs at American University's School of International Service in Washington, D.C. She has published extensively on modern and contemporary China, including *Mao's War against Nature* (Cambridge, 2001), an account of China's political campaigns to conquer the natural world.

SHEN Xiaohui is a member on the Chinese National Committee of Man and Biosphere, deputy editor-in chief of *Green China* magazine, and executive council member of Friends of Nature. He has done extensive research on forests, wetlands, and deserts in China. In 2000 Shen joined the writing project on the Environment and Development in China sponsored by the Chinese Academy of Social Sciences.

SHI Pengxiang is a staff member of the Greenpeace office in Beijing.

Tamara STARK is a staff member of the Greenpeace office in Beijing.

SUN Jun is the director of Green Cross, an environmental NGO based in Beijing. He is a council member of the Beijing Sustainable Development Educational Association under the auspices of UNESCO and a project manager with the Department of Environment and Sustainable Development of the China Academy of Social Sciences. He has long been committed to studying and solving rural environmental problems and is among the first to bring the practice of refuse collection and processing to China's rural areas.

TAO Chuanjin obtained his PhD in Sociology from Renmin University of China in 2002. From 2002 to 2004 he did post-doctoral research in the School of Public Policy and Management of Tsinghua University. During this time he worked as a visiting scholar for a year at the University of Chicago. Currently he is an associate professor with the

Institute of Social Development and Public Policy of Beijing Normal University. He has published dozens of articles and a book titled *Environmental Administration: Starting from Communities*.

WANG Peng graduated from the Department of Sociology of Peking University. He is an active member and volunteer with Friends of Nature. He is committed to environmental protection in China and has participated in many campaigns to conserve the environment.

WU Jian is an instructor of environmental economics at Renmin University of China. Her major field of research is economic policies on environmental and natural resources management. She has participated in four international research projects on pollution emission-exchange rights. Recently she has been engaged in various research projects sponsored by the China Council for International Cooperation on Environment and Development (CCICED), the World Bank, the Asian Development Bank, and Environmental Defense.

WU Zhijiao is a graduate student from the Chinese University of Political Science and Law and a member of the Center for Legal Assistance to Pollution Victims in China.

ZHANG Ying is an instructor with Beijing Forestry University. Since 1994 she has been engaged in research on forestry resources accounting. After completing her doctorial dissertation in 1999 she did post-doctoral research at Korea University. Among her research topics is how to account for the value of diverse forest types, fauna, and flora in China.

ZHAO Wengeng is an editor from *Water Supply and Distribution* magazine. He is a member of Friends of Nature.

INTRODUCTION

You hold in your hands a remarkable volume. This edition of *The China Environment Yearbook*, or "Green Book" as it is affectionately called, is the second in a series of informative annual records written, commissioned, produced, and edited by Friends of Nature, China's premier environmental citizens group. Daring in its perspectives and investigative reach, rich in serious research, documentation, and rigorously compiled statistical information, this book provides a window on the year's debates and events as they have affected China's struggles toward a more just and sustainable model of development. Many of the case studies are based on first-hand knowledge and long-term determination and commitment, such as Huo Daishan's work on behalf of the Huai River, so readers may understand the human stories behind the numbers and can admire and appreciate the love of many of China's environmentalists for their country.

Readers of this volume will be introduced to key concepts and controversies, such as innovations in the trading of water rights between regions and cities and the furor over the film industry's environmental carelessness. They will read many courageous essays: One of these examines policies of fencing Inner Mongolian grasslands in a way that contradicts local culture and ecology; another questions the wisdom of the South-to-North water transfer scheme in the upper Yangzi (and of a potentially even more ecologically intrusive mega-project called the Shuotian Grand Canal Project). We learn about abuses of forest rights "reform" measures which have unexpectedly promoted clear-cutting of old growth forests, mono-cropping of eucalyptus, and planting of genetically modified poplars. Another article exposes failures in government efforts to clean up some of China's most heavily polluted waterways. There are candid accounts of new levels of environmental degradation in rural areas and of the difficulties encountered in China's effort to produce a "green GDP" that would accurately reflect the costs of natural resource extraction and pollution. (The effort to account for the real environmental costs of economic growth, both in terms of depletion of "natural capital" and costs of pollution, is highly subject to manipulation of statistical data, and China has found itself to be no exception.) Other hard-hitting articles describe China's role in the global trade in illegal logging, publicize the problem of "cancer villages," detail cases of village protest against polluting factories, and make clear the seriousness of problems with widespread groundwater contamination and lack of access to safe drinking water.

While these issues are similar to those encountered in other places in the world, many of these articles are written with a tone and orientation that is uniquely Chinese. Why, for example, does the volume begin with a wise contribution from Hu Kanping, the most senior contributor? In China, deferring to status is a characteristic way of providing both weight and political protection to the project as a whole. Why is senior SEPA official Pan Yue's discussion of ideology so important and courageous? As Chinese readers would know, Chinese politics are guided by ideological debates on such issues as the definition of "Socialism with Chinese characteristics" (a phrase which allows China to embark on its current market-oriented phase of economic growth without renouncing the leadership of the Communist Party). Pan's nuanced effort to integrate principles of sustainability into the nation's guiding tenets has the potential for far-reaching consequences, and is articulated at some political risk. Why does the description of the Wushan model, which promotes sustainability at the village level, remind a reader of the Mao-era campaigns to "learn from" Dazhai? Perhaps China is still culturally enamored of a form of social change based on top-down moral suasion, imitation, and self-cultivation. These are just a few examples of how this volume's uniquely Chinese voice is part of its exceptional value as a primary source.

We should not forget how remarkable it is that such a volume can be produced in China. A mere thirty years ago some of the senior contributors to this effort were in exile in the countryside. College entrance examinations were not reinstated until 1977, and Mao's nature-conquest ideology was still creating widespread suffering for both people and the land against which they were required to struggle. Since then, an ideology of capitalist consumption has supplanted Maoism to accelerate China's environmental degradation to the point that, in the words of a recent *New York Times* series, it is "Choking on Growth." China's environmental problems have become a world-notorious threat to the Chinese people's aspirations for a better life.

That a citizen's group like Friends of Nature can produce such a series is the more remarkable given this recent history. Chinese citizens groups tread upon uncertain ground. What is considered an acceptable topic for debate one year may become "inconvenient," sensitive, or even dangerous the next. All citizens groups are expected to register with a governmental institution (often known as their "mother-in-law") and avoid behavior that might cause them to be shut down. The political climate may shift, sometimes unpredictably. But when "cool winds" blow in one part of the country, "warm winds" may prevail elsewhere; a cold snap is eventually followed by a thaw. Fortunately, China's government has come to appreciate the value of its environmental citizens groups as essential allies in the national effort to move China toward sustainable development.

The perspectives expressed in this volume are thus priceless, and this series of volumes, taken together, will provide a record of the victories and setbacks of China's people as they struggle to gain better information concerning pollution, deforestation, and ecological degradation; to hold their local and national leaders accountable for the environmental performance of the institutions that affect their communities; to gain a greater say in decisions concerning major infrastructure development projects; and to freely participate in debate over the values and direction of the nation. The publication of Volume Two of the *China Environmental Yearbook* is a testament to the national concern about environmental issues as well as the increased space for debate in comparison with a few short decades ago.

It is an honor for me to write this short note of introduction for English audiences, and my co-series editor Guobin Yang and I look forward to greater participation in the project in coming years. However, in closing I wish to underline that the book is a Chinese effort and as such can be regarded as a primary resource. Its value for researchers and policymakers should endure for many years to come.

Judith Shapiro
American University

GENERAL REPORT

CHAPTER ONE

STANDING AT A NEW VANTAGE POINT—CHINA'S ENVIRONMENT IN 2006, THE FIRST YEAR OF THE ELEVENTH FIVE-YEAR PLAN

Hu Kanping

2006 is the first year of China's Eleventh Five-Year Plan (2006–2010). This Five-Year Plan marks an historical change in environmental policy advocated by the Chinese government.

In 2006, China's economy continued to grow rapidly. This growth follows the so-called "extensive growth model," meaning that it demands vast resource inputs and generates huge gross outputs, particularly in heavy industry. Unfortunately, this model generates severe pollution. Despite some signals that development is moving in a different direction, the current momentum seems irresistible.

Meanwhile, ever-worsening environmental and resource crises threaten people's livelihoods and social stability.

Should the extensive growth model continue to guide development, growth will be achieved at the expense of energy conservation and environmental health and improvements. This direction steers China towards disaster instead of prosperity. To avoid this outcome, rational resource utilization and effective ecological protection must be integrated into economic policies since this balanced system forms the basis of a well-balanced society. Environmental measures in the Eleventh Five-Year Plan form a transition away from the reckless application of extensive growth to a more balanced system where environmental needs and constraints enter into policymaking and decision making processes. In this way the Eleventh Five-Year Plan period is crucial for building a harmonious society[1] based on rational development policies.

[1] Translator's note: The phrase "harmonious society" may conjure images of people getting along socially, but its use in China has a subtle, rich, and deeply layered meaning. The words "harmonious society" imply and suggest a wide range of related actions and conditions: building cooperation and coordination among individuals and organizations; successfully resolving social tensions and seeking balance between various social and

Environmental protection is a key element in achieving comprehensive, coordinated and sustainable development that puts people first, but within the broader and interdependent system of human, social, and environmental needs and constraints.

The economic and social planning in the Eleventh Five-Year lays out the major goals that must be achieved by 2010 regarding environmental protection. It is the first time that goals for energy conservation, pollution control, land protection, and forest cultivation have been stated in indexes used to evaluate economic growth and social development. Environmental issues, such as resource protection, have become the ties that bind our national will and drive it to better balance the needs of individuals, nations we share resources with, and generations yet to come. Clearly these goals can be realized, but they will depend on implementation efforts rather than administrative orders.

The GDP growth rate has been presented as a "soft" index—that is, an index that is a hoped-for target rather than a fixed goal—for the first time by the Chinese central government. In the past, "the overriding importance of development" was mistakenly equated to "the overriding importance of economic growth." As a key index of economic growth, GDP has been frantically pursued by local government officials. Such a trend is hard to change quickly, but some changes have begun to take shape.

On this heavily polluted land, changes are taking place and green thinking is beginning to take root across the nation. In 2006, the Chinese government took it upon itself to move from a model of economic growth at the expense of the environment to optimization of economic growth within the context of environmental protection. To this end, the Sixth National Environmental Protection Conference held on April 17 and 18, 2006, served as a milestone in the history of China's environmental protection efforts. Premier Wen Jiabao called for "three changes" at the conference: first, making the change from a single-minded pursuit of economic growth while neglecting environmental protection to emphasizing a balanced development strategy that

economic players; and avoiding and suppressing conflict while engendering peacefulness. This phrase suggests a more fluid kind of egalitarianism; equality is not a static state, but rather a process. The Chinese mean these things in a comprehensive way: such values should be at work within the individual, families, the society, and even the international sphere. This rhetoric draws upon Confucian values, and to some degree reframes the dialectical values of doctrinaire communism.

considers both environmental protection together with economic growth; second, moving away from a system where environmental protection lags behind economic growth to a system that synchronizes economic growth and environmental protection; and third, changing from a policy system that leans heavily on administrative leverage to a comprehensive system using legal, economic, and technical tools alongside administrative measures to solve environmental problems. The three changes are interdependent; seeing the need for these changes and implementing them as an integrated whole marks a significant change in strategic policymaking for the state. In fact, Zhou Shengxian, Minister of the State Environmental Protection Administration (SEPA), calls this shift an "historic transformation.'"

Unlike Western countries, environmental protection in China was government-dominated from the very beginning. However, business and scientific communities and NGOs are becoming major forces for environmental protection as globalization quickly spreads. Domestic conditions support their efforts as governmental functions continue to transform, markets mature, and civil society grows more fully developed and robust. China's current environmental protection situation is characterized by multiplicity, featuring "governmental guidance, business leadership, public participation, scientific and technological support, along with legal and market regulation." All forces of environmental protection, domestic and foreign, governmental and non-governmental, are now standing at a fresh starting point and are seeking new ways of considering development along economic and social lines.

2006 witnessed fundamental environmental changes in China. That year saw remarkable achievements fulfilled; however, restraining factors still existed. Conflicts and tensions between various forces have created rewarding ups and disappointing downs on the road to achieving greater environmental protection.

I. *The Songhua River Toxic Spill Crisis as a Catalyst in China's Environmental Protection Movement*

China's environmental protection efforts in 2006 took shape under the cloud of the Songhua Toxic Spill Crisis, which occurred in November 2005. Upon taking office, Zhou Shengxian, the newly appointed Minister of the State Environmental Protection Administration, went to Heilongjiang Province to lead the fight against pollution in the Songhua

River. Shortly thereafter, he made a pledge to his staff, saying, "I will fight to guarantee that scientific insights and rational analysis will be applied in the service of environmental protection."[2] Before the Chinese New Year, the Chinese government declared victory in the protracted battle involving the condition of the river. This announcement gave those living along the Songhua River some peace of mind that their interests were safeguarded. Chinese New Year celebrations in the area carried a special sense of joy because of this.

Just as a coin has two sides, the Songhua River Spill Crisis comes with a silver lining. This crisis shocked people into an awareness that environmental protection will be a long, hard road to travel. The multi-dimensional nature of the crisis touched on many related points that underscored the scope of the challenges lying head. The threats in this case included the potential dangers posed to individuals' health, social stability, ecological safety, economic development, and even international relations, as the river flows along the border between China and Russia. The crisis was also important because it gave officials working in departments and bureaus in charge of environmental protection a strong sense of responsibility.

Out of the crisis a new spirit was born, which Zhou Shengxian described as "being faithful to one's duty, working whole-heartedly for the welfare of the people, upholding scientific analysis, showing no fear in the face of difficulties, and being selfless and cooperative in dealing with emergencies."

II. *Energy Conservation and Environmental Protection as Hard Indexes*

During the National People's Congress (NPC) and Chinese People's Political Consultative Conference (CPPCC) sessions held in March 2006, Premier Wen Jiabao told journalists at a press conference, "It is true that we are now faced with the very serious problem of environmental pollution that plagues China's development. Even now, we fail to see a proper solution to this problem. Talking about the Tenth Five-Year Plan, I can tell you that we have been able to meet the original goals in most of the categories. At the same time, to be honest with you, we

[2] Nie Xiaoyang, "New Chief Took Office: Swearing to Be a Fighter, a Hard Worker and a Protector," *Oriental Outlook Weekly* 12 (2005).

failed to deliver on the targets concerning environmental protection."[3] He was referring specifically to targets aimed at improving the urban environment and cutting the total discharge of major pollutants by 10 percent from 2000 levels as defined in the Tenth Five-Year Plan.

Most Chinese are aware that environmental protection is the weakest link in China's development. However, environmental protection agencies should not take all the blame for these shortcomings. Reducing the discharge of major pollutants seems very unlikely as long as the extensive economic growth model remains in place. Minister Zhou of SEPA stated flatly that this model requires that GDP growth targets are achieved by "burning up resources and leaving behind pollution." Pan Yue, Vice Minister of SEPA pointed out:

> China's GDP growth is generated by high energy consumption and high pollution. 90% of air pollution is caused by heavy industry and 70% from power generation. Both are increasing by 50% per year. This creates a voracious demand for coal. Coal burning accounts for 85% of overall energy consumption. If this situation continues, environmental issues will become more severe, and more social problems will occur. Although a policy concerning desulphurization has been made, 95% of power plants have not installed or do not use desulphurization equipment. It's beyond our power to do anything about this. The fundamental economic mechanism is the problem.[4]

The outline of the Eleventh Five-Year Plan for Economic and Social Development (2006–2010) passed at the NPC and CPPCC sessions sets forth environmental protection goals for the next five years. By 2010, while the national economy will continue on its trajectory of rapid, stable growth, the environmental quality of key regions and cities will be improved, and the trend toward ecological degradation will be brought under control. Specifically, energy consumption per unit of GDP will decrease by 20 percent from 2005 levels; the total discharge of major pollutants will be reduced by 10 percent. Furthermore, forest coverage will increase from the current 18.2 percent of land area to 20 percent. These indexes are "hard" in the sense that they are obligatory, while the GDP growth rate is a relatively "soft" index, meaning that the target rate is only an expected value. For the first time in China's development,

[3] Xinhuanet, March 14, 2006.

[4] Pan Yue, "Economic Development and Environmental Protection," Ministry of Environmental Protection of the People's Republic of China, http://www.zhb.gov.cn/hjyw/200612/t20061220_97538.htm.

environmental protection indexes are put on equal footingwith economic growth indexes, and in some cases are given higher priority.

Premier Wen Jiabao said that although it is difficult to reach this goal, we have the confidence and determination needed to achieve it.

Table 1.1: Major economic development and environmental protection indicators during the Eleventh Five-Year Plan

Economic Variables	Leading Indicators	2005	2010	Annual Growth Rate (%)	Attributes
GDP (trillions of RMB)		18.2	26.1	7.5	Expected
Reduction of Energy Consumption per unit of GDP (%)	20				Obligatory
Reduction of Water Consumption per Industrial Added Value (%)	30				Obligatory
Total Cultivated Land (billion hectares)		.122	.12	−.03	Obligatory
Reduction of Major Pollutants Discharge Volume (%)	10				Obligatory
Growth of Forest Coverage (%)	1.8	18.2	20		Obligatory

During the National Conference of Directors of Environmental Protection Bureaus, Minister Zhou Xiansheng of SEPA put forward the target of cutting the total discharge of major pollutants by 2 percent before the end of 2006. Early indications suggest that he may have been overly ambitious in setting this goal, even in the face of environmental conditions that call for setting the goal even higher. This point is addressed below.

III. *Historical Change*

On April 17, 2006, the residents of Beijing awoke to a city nearly covered in sand after a particularly intense sandstorm. Open ground, houses, automobiles, and even tree leaves had a heavy layer of sand on them. The storm deposited three hundred and thirty thousandtons of sand on the capital. With Beijing's population of fifteen million people, this amounts to twenty kilograms per person. Premier Wen Jiabao attended the Sixth National Environmental Protection Conference during the sandstorm and delivered a very important speech. In this speech, he made it clear for the first time that doing environmental work must now incorporate the three changes defined above. He emphasized that we must respect the law of both economics *and* nature in order to better protect the environment.

This historic change demands that we set aside our reactive mode of action, which pushes us to treat pollution and work for environmental improvements only after damage has been done. Instead, we must be proactive and far-sighted; preventive measures should be at the core of our activities. Of course, this change of working style will happen gradually rather than overnight. But it must begin now. Putting this "historical change" into effect helps to give people the confidence needed to pursue and fulfill environmental protection actions.

IV. *A Difficult Beginning*

Although statistics for the entirety of 2006 are not yet available, data from the first half of the year do allow for preliminary analysis of the progress made at the beginning of the Eleventh Five-Year Plan. Unfortunately, these early returns are disappointing.

The Eleventh Five-Year Plan set a goal of 20 percent reduction in energy consumption per unit GDP in 2006. Based on this, the *Government Work Report* set a goal of a 4 percent reduction for the year. Through June, however, this index rose by 0.8 percent since the close of 2005.

The Plan also aimed to cut the total discharge of major pollutants by 10 percent over the course of the five-year period of the plan, from 2006–2010. Specifically, Chemical Oxygen Demand (COD) has been slated to fall to 12.73 million tons, after starting at 14.14 million tons in 2005, and sulfur dioxide has been targeted to decline to 22.94 million tons from a starting point of 25.49 million tons in 2005. The goal for 2006 was a 2 percent reduction in total discharge of major pollutants. However, the data reveal that in the first half of 2006, COD emissions were 6.896 million tons, an increase of 3.7 percent over the same period in the previous year, and sulfur dioxide emissions were 12.46 million tons, a 4.2 percent year-on-year jump. Based on this, we can clearly see that the facts on the ground trump the tough talk found in recent public pronouncements and policy statements.

SEPA has put pollution discharge reductions at the top of its agenda for 2006, requiring every city, county, and polluting enterprise to meet its target. In May 2006, acting under the aegis of the State Council, SEPA signed agreements on sulfur dioxide emission reduction with Shandong and six other provinces, as well as China Huaneng Group and five other major power suppliers. In July, SEPA signed another set of similar agreements with Hebei and eight other provinces. The remaining provinces committed to these reductions in August.

Some analysts have pointed out that the failure to meet these targets developed as a result of a number of new pollution-emitting projects coming online during the first half of 2006. These projects had been approved two or three years earlier. By coming online within two quarters of each other, a "bump" in pollution output could not be avoided. This almost certainly aggravated a difficult situation.

Note that starting in January 2004, power suppliers and other major emitters of sulfur dioxide began installing Flue Gas Desulphurization (FGD) equipment on a large scale. By the end of 2005, fifty-five million–kilowatt-sized coal-burning units had installed FGD equipment, accounting for one-tenth of the total coal-burning power generation units of this scale in China.

The State Council will conduct midterm and final assessments on the implementation of pollution reduction plans in all provinces from 2008 to 2010. In the past, agreements between the State Council and the provinces concerned the management of such efforts' final stages without specifying annual reduction targets. The change in the new agreement caught the provinces unprepared to meet annual goals. This is one of the main reasons why the provinces did not actively engage in reduction efforts during 2006.

V. *The Cost of Growth*

The ordinarily "soft" targets concerning energy conservation and environmental protection cannot become obligatory, "hard" targets overnight, nor will GDP, the original "hard" index, become "soft" right away. Even though GDP growth targets are on a downward trend, actual GDP has diverged from the target rate. During the first quarter of 2006, China's GDP growth rate amounted to 10.2 percent, but for the first half of 2006, the growth rate amounted to 10.9 percent. This indicates that GDP growth continues to trend upward. Furthermore, the 10.9 percent growth rate exceeds the growth rate target of 8.0 percent by a wide margin. The National Development and Reform Commission (NDRC) projects the GDP growth rate to be no less than 10.2 percent for the entire year.

Given recent experience, the final growth rate for 2006 may be even higher. The National Bureau of Statistics in China issued a report in 2005, which shows that China's GDP growth rate was 9.9 percent over the previous year, while the average growth rate of thirty-one provinces and municipalities was actually 12.39 percent. According to a report issued by the NDRC, all thirty-one provinces and municipalities (excluding Hong Kong, Macao, and Taiwan) maintained double-digit growth rates through the first two quarters of 2006; twenty-threeof them achieved growth rates of over 12 percent. If China's GDP growth rate was calculated according to the announced growth rate of all provinces and municipalities, the actual growth rate should be closer to the 12.39 percent that they collectively reported rather than the much lower figure of 9.9 percent.

While many economists at home and abroad marvel at this "tremendous achievement," many of us feel something other than unalloyed pride when we see the gigantic costs to the environment exacted by this breakneck pace of economic growth.

Many experts pointed out long ago that the existing GDP calculating method could not truly reflect the full scope of development since orthodox economic accounting omits environmental and ecological factors.[5] However, opinions vary widely as to the scale and scope of these two factors. According to a report by the World Bank and some experts from the Chinese Academy of Sciences, economic losses due to environmental damage amounted to 15 percent of GDP in 2003. A survey of 973 research projects on desertification, entitled *Deserts and*

[5] Translator's note: Various scholars and organizations have called for a push to use "triple bottom line" accounting methods to better account for growth and a more comprehensive definition of development. See John Elkington's 1994 paper, "Towards the sustainable corporation: Win-win-win business strategies for sustainable development," which appeared in *California Management Review*. A number of economists have called for a similar approach dubbed "true cost economics." Kenneth Boulding, Robert Costanza, and Herman Daly, among others, have led efforts to legitimize and formalize this style of economic analysis.

Desertification in China, shows that in recent years the average annual economic loss caused by ecological degradation and environmental pollution amounted to roughly 14 percent of GDP. The two results seem to reinforce each other. However, the reader should keep in mind that the aforementioned data are not exact.

An official report issued in 2006 offers support for the call to adjust GDP calculations, or at least to consider a variety of costs overlooked in the official GDP calculation. This report considers a wide variety of ecological costs, including resource consumption. Specifically, the report names such costs as: damage to land resources; depletion of mineral resources; destruction and overuse of forest resources; pollution of water resources; and damage and overuse of fishing resources. The report identifies various forms of pollution, including contamination of underground water supplies, soil pollution, and eight other major forms of pollution. Taken together, these costs amounted to 511.8 billion RMB, or 3.05 percent of the GDP, in 2004. This calculation is the basis of a second calculation that estimates that a one-time investment of about 1,080 billion RMB, or about 6.8 percent of GDP in 2004, would be needed to properly treat or dispose of pollutants emitted in 2004. This puts the costs of pollution at nearly 10 percent of GDP, or roughly the amount by which GDP grew in 2004.

These data are derived from the *China National Green Accounting Report for 2004*, issued jointly with SEPA in September 2006. The attempt to produce a "Green GDP" based on a broader accounting of economic and environmental data was initiated in 2004. The goal was to establish a framework of integrated environmental and economic accounting in the next three to five years.

In November 2006, the ten provinces and municipalities that conducted this experiment expressed their unwillingness to continue using green GDP accounting methods in their provinces and municipalities. Clearly, the goal of implementing a green accounting system will encounter stiff resistance from some quarters.[6]

[6] Translator's note: Efforts to revise GDP in the West have proven equally fraught. For instance, few economists, policymakers, or business journalists report on measures such as the Genuine Progress Indicator (GPI). Follow this link for more information on this measure: http://www.rprogress.org/.

VI. *Competition between Two "Protections"*

In 2006, the American author and journalist Thomas L. Friedman published two books on globalization, *The World Is Flat: A Brief History of the Twenty-First Century* and *The Lexus and the Olive Tree*. Both went on to become bestsellers while garnering considerable attention in the media. The books even had a substantial impact among opinion leaders in China. During a speech that he delivered in China, Friedman posited his "green cat" theory as a foil to Deng Xiaoping's more famous "white cat, black cat" metaphor.[7] Friedman claims that the cat had better be "green" or it may die before it catches any mice.[8]

Environmental protection can lead to productivity improvements; linking economic benefit to environmental improvements is at the core of the argument for pursuing green policies. However, traditional notions dictate that spending on environmental measures raises the costs of production and may hamper efficiency. Because of this, many officials and business leaders do not believe that a "green cat" is needed. Carrying the metaphor to its logical end, it is as if they find it acceptable that the cat catch a dead or even poisonous mouse if it does not catch a living one.

Beyond this broader philosophical divide lies the schism between the central government and local governments, as well as the split between superiors and their subordinates at the local level. In many ways, these tensions operate much like a game of cat and mouse.

A report entitled *Environmental Governance in China*, completed by the China Council for International Cooperation on the Environment and Development in 2006, pins China's continuing energy and environmental problems on its lack of effectiveness in implementation and oversight of policy. In short, the problem is governance, or lack thereof.

The core of the problem is that central government policies are not effectively implemented at different governmental levels. Policies are modified or diminished throughout this process; in the extreme almost nothing of the original policy is left intact. This has a major impact

---

[7] Translator's note: This metaphor references a well known speech given by the General Secretary to the Party Secretariat, in which he said: "No matter if it is a white cat or a black cat; as long as it can catch mice, it is a good cat." This is also a proverb from Sichuan.

[8] The basic ideas in this speech come from Thomas L. Friedman, "Bring in the Green Cat," *New York Times*, November 15, 2006.

on economic, and therefore environmental, policies. For instance, the central government, which considers social interests in the broadest terms, intends to shift from the extensive growth pattern as the basis of economic policy. From its "bird's eye" view, it can see that this will help to maintain social stability and sustainable economic development. Unfortunately, the interests of local governments may run counter to these aims; therefore, local government leaders and bureaucrats may work against central government policy. As it stands, local officials are not penalized for neglecting central government policies, especially those regarding the adjustment of economic structure. Since the extensive economic growth model continues to pay dividends—and since there are no clear political costs to continuing on this path—the best option for local leaders is to stay the course. This allows them to build their political profile, increase local economic benefits, and show "outstanding achievements" in fostering economic growth.

Some analysts point out that under the current system, different localities function within localized financial systems, which means that the living standards and incomes of local officials are decided by the financial situation of the local government. Under such circumstances, local governments are likely to pursue a high rate of economic growth in order to improve the financial condition in their locality. This is why many local governments choose to hold fast to the increasingly dubious practice of "pollute first, treat later."

Leaders push for the expansion of investment scale, especially foreign capital. This process often moves forward with little or no planning. Investment often goes into projects with high energy consumption demands. Worse, many of these projects output vast amounts of pollution. Environmental capacity, resource constraints, and even basic financial considerations are overlooked.

Protectionism in different localities is the main cause of ineffective enforcement of central government laws and regulations. Local governments create and implement all manner of local policies that counter those from the central government, including the relaxation of environmental protection standards as a way of attracting foreign investment. Some local governments even go so far as to grant special waivers that allow some polluting firms to avoid on-the-spot inspections by environmental officials. Some bargains go further still, granting those firms that flout regulations a "green light" to do so. Some local governments regard environmental protection bureaus as ornamental; token pressure is applied to supposedly keep local environmental officials in line. Local leaders see such efforts as impediments to the all-important goal of economic growth. Caught in the middle of this dilemma, many directors of local environmental protection bureaus "choose" to retire rather than continue in a role that has almost no support or impact. Those that stay on may deliberately refrain from using their authority and power. Some may go so far as to serve as patrons and protectors of local enterprises, in spite of the central government's efforts to curtail offenses by these firms. In essence, local environmental protection bureaus sometimes choose between protecting the environment and protecting polluters.

Nowhere are such problems more visible than in Shanxi Province, which ranks first in terms of pollution in China. Total discharge of major pollutants far exceeds environmental capacity. In this province the environment has been stretched to its limits. According to a survey conducted by the Shanxi Environmental Protection Bureau in November 2006, 93.31 percent of the people surveyed believed that economic growth and environmental protection should go hand-in-hand, while more than 91.95 percent of local government officials worried that increased environmental protection efforts may hinder economic growth. Note that only 6.51 percent of local government officials disagreed with this view.

The central government is now well aware of the seriousness of these problems. At the Central Economic Work Conference held in Beijing at the end of 2006, the questionable policy of single-mindedly pursuing rapid economic growth was replaced by a policy based on the need to realize a higher quality of development. In this view, "quality" supersedes sheer "speed."

The conference developed and advanced four points of focus regarding economic policy in 2007: adjusting the economic structure by changing the economic growth pattern from an *extensive* system to an *intensive* system;[9] strengthening resource conservation and environmental protection; promoting reform, liberalization, self-reliance, and innovation, including upgrades in skill levels and productive technologies; and

[9] Translator's note: *Extensive growth* refers to a system of economic growth built around expanding output and using vast resource inputs to achieve this outcome; this model is described and critiqued elsewhere in this essay. On the other hand, *intensive growth* refers to a system where efficiency is emphasized. Furthermore, this system of growth focuses on production that moves up the value chain and seeks to produce goods and services with higher value-added. Note that this system of growth differs sharply from the current regime in place in China that emphasizes low-cost mass production.

accelerating social development while raising living standards across the board. Conferees also mapped out the requirements needed to implement the "three areas of coordination": speed, quality, and efficiency; consumption, investment, and exports; and population, resource use, and the environment. This language reflects the central government's increased attention to quality of development and efficiency of processes in achieving economic development. Clearly, the central government is determined to address major problems affecting economic and social development.

However, it remains to be seen if the central government's policies will be supported by local governments.

VII. *Lingering Risks*

In 2006, the environmental risks that China faced did not disappear. For example, the total discharge of major pollutants exceeded environmental capacity. Water, air, and soil pollution worsened, and solid waste, car exhaust, and various forms of organic pollution increased. In recent years, pollution cases concerning collectives increased by 29 percent per year. Fifty-one thousand cases related to environmental pollution were filed during 2005. Emergency pollution events occurred once every two days on average in 2006. From January to June, SEPA settled eighty-six emergency environmental incidents, which already exceeds the total of seventy-six cases in 2005. Most of these cases were related to water pollution—for example, the lead poisoning incident in Hui County, Gansu Province, and the arsenide[10] pollution problem in Yueyang County, Hunan province. These emergency incidents occurred frequently in China—one occurs about every other day—and individually and collectively they pose serious social and environmental risks.[11]

The public's feelings regarding environmental quality arise from deep feelings in their hearts. Their reactions and statements can be profound and blunt. A story from spring of 2006 provides a useful example. A woman in Guangzhou jumped into a river in an attempt to commit suicide. After being saved by passers-by with a ladder, she said, "The

[10] Translator's note: Arsenide is also known as gallium arsenide. It is a compound made up of gallium and arsenic.

[11] Some of the increase in such incidents is due to better monitoring and reportage, but it is unknown to what degree this affects these statistics.

water is so smelly and dirty that to be soaked in it is more unbearable than death itself."

There is a popular folk song in Shanxi Province, which sings of the clear water and blue sky there. However, in some villages, the villagers have changed the lyrics to something like this:

> Black smoke is clouding the sky
> Smelly air rushes into our houses
> And we are forced to close our windows
> And doors all day long.

In Henan Province, a popular ballad echoes this sentiment:

> The sun is like the moon
> A fine day is like a cloudy day
> And our nostrils are like chimneys.

One may argue that these are only isolated cases. But the following data and facts can help us to see China's current environmental situation more broadly and clearly.

The first group of data shows that 70 percent of the rivers in China are polluted, and 90 percent of the rivers flowing through cities are "seriously polluted." According to *The Monthly Report on the Quality of Drinking Water Sources in 113 Environmentally Protected Major Cities*, issued by the Chinese Environmental Monitoring Centre in June 2006, the water quality of sixteen—or 14 percent of the total—cities completely failed to meet minimum drinking water standards. Even more troubling, seventy-four drinking water sources—or 20 percent of the total—failed to meet minimum standards. In total, 527 million tons of water, or 32 percent of fresh water volume, failed to meet minimum standards. In the rural areas, 34 percent of farmers (over three hundred million people) do not have reliable access to clean drinking water. The *2006 Survey on Environmental Safety in China* revealed that of more than seven thousand five hundred projects in the petroleum and chemical industries, 1,354 of them are constructed on the banks of rivers, lakes, reservoirs, and seas. Two hundred and eighty of them operate within ten kilometers of protected drinking water sources, and 2,489 of them operate near cities or densely populated areas. If accidents involving pollutants occur, the consequences will be severe. One only need look at the Songhua River Crisis to get some sense of the potential harm we could witness.

Water pollution is not the only problem; China has the highest incidence of lung diseases in the world. More than four hundred million urban residents rarely breathe fresh air. Twenty percent of children suffer some degree of lead poisoning through inhalation. A report issued by the World Bank in 2006 said that in big Asian cities, one person dies of air pollution every minute. Xi'an, the historic capital of China, and Beijing, the current capital of China, experience some of the worst air pollution not only in Asia but in the world.

Problems do not end with water and air pollution. Contamination of food supplies has been on ongoing issue. In 2006, China witnessed new kinds of food pollution: Sudan Red 1[12] and Malachite Green[13] have turned up repeatedly in food supplies. According to a research report on the environmental and health conditions in China issued by the Chinese Academy of Sciences, 75 percent of chronic diseases are related to pollutants ingested through food.

Premier Wen Jiabao once said: "Our goal is to make sure that all people can have clean water to drink, fresh air to breathe and safe food to eat." These basic elements for the existence of human beings did not cause any concern for our ancestors who lived thousands of years ago, but now we have to worry about what we drink, breathe, and eat. So much for all activity being equated to progress. And so much for equating a reckless model of economic growth with development.

Other data and facts prove equally shocking. In China, the urban garbage treatment rate is less than 20 percent; over one hundred and fifty million tons of garbage in rural areas simply pile up in open areas. There are 3.56 million square kilometers afflicted with water and soil erosion; desertification affects 1.74 million square kilometers. One-third of China's land has been damaged by acid rain. Ninety percent of natural grasslands have suffered some degree of degradation. Moreover, the number of endangered and threatened species is increasing; genetic resources are in decline. Biodiversity of both flora and fauna

[12] Translator's note: Sudan Red 1, also known as Yellow Solvent 14, is an industrial, red, oil-soluble dye, used legally to color a number of non-food products such as shoe and floor polishes, oil, waxes, and solvents. It has been shown to cause cancer in rats, and consequently its use in any food or food product is not allowed. Despite this, the dye has repeatedly found its way into food products, including chili powder; the dye's red color enhances marketability of this and other condiments. As a result, traces of the dye are found in other products that include chili powder as an ingredient, including blends of spices, sauces, various spicy foods, and so on.

[13] Translator's note: Malachite Green is a synthetic dye used to color silk, wool, hemp, leather, cotton, and paper. Malachite Green has been traditionally used to treat fungal infections on fish eggs. It is not approved for human consumption. Reports published in 2006 identified Malachite Green in fish meant for human consumption in China.

has been hit hard. At present, 15–20 percent of all plants in China are endangered or threatened.

Shoreline waters are widely and severely polluted. Water near estuaries and cities has suffered the most: 59.7 percent of the water is at Level III or below Level III.[14] In 2006, the United Nations announced the Yellow and Yangtze River Estuaries as "dead zones" since fish, shrimp and other shellfish, and even seaweed and other plant life cannot live in these places.

Pollution causes 70 percent of cancer deaths per year. One out of every ten couples in big cities cannot bear children because of pollution; the incidence of infertility among women in Shenzhen City attributable to environmental pollution has recently increased by almost 20 percent.

The large volumes of industrial discharge have been widespread as well as severe; clean rivers and lakes have become a rarity. While we are ambitiously working as "the world's factory," another part of the international community has classified China as "the world's dumping ground."

Efforts Made by All Parties

On December 22, 2006, the Northeastern Supervision Center for Environmental Protection under SEPA began operating in Shenyang. Soon after, supervision centers for environmental protection started operating in five major regions as described below.

The large number of reported pollution problems and events can ultimately be attributed to violations and disregard for environmental law and regulations. Research shows that only about 30–40 percent of county-level projects requiring environmental impact assessments (EIAs) complete such assessments. Less than 50 percent of projects in industrial parks submit to an EIA. Of coal-fired power plants going into

[14] The Surface Water Quality Standard passed by SEPA in 1988 rates water in six categories. Level I is suitable for river source water in nature preserves. Level II is suitable for drinking water without treatment, swimming, Grade I Protected Areas, and rare and sensitive aquatic life. Level III is suitable for drinking water with some treatment, swimming, Grade II Protected Areas, and ordinary and unprotected aquatic life. Level IV is suitable for non-body contact recreational use and industrial use. Level V is suitable for agricultural and landscaping. Below Level V (Level V-) is completely toxic and unusable. Note that Grade I protected areas include river source areas and national-level nature preserves.

production from January to June 2006, which total 32 million kilowatts in output, only half are equipped with desulphurization technologies. An investigation conducted by the relevant departments examining recent construction projects of one hundred million RMB or larger across eight provinces and regions shows that about 40 percent of the projects have violated at least one law or regulation. Violators ignore or side step rules and regulations on land requisition, environmental assessments, and various kinds of "required" inspections. Unfortunately, SEPA does not have the capacity or authority to sanction or remove local government officials who willfully allow or actively promote such problems in order to realize short-term economic benefits. Furthermore, SEPA cannot directly intervene in the management of local environmental protection agencies. Given this, the lack of compliance with government policies is no surprise.

As the saying goes: "A severe ailment needs strong medicine." Similarly, to solve a nagging problem, decisive measures should be adopted. On July 31, SEPA announced that it had set up eleven local environmental protection supervision and enforcement offices. They would work with five environmental monitoring centers in East China, South China, Northwest China, Southwest China, and Northeast China, and six monitoring centers tracking nuclear and radiation security in Shanghai, Guangdong, Sichuan, North China, Northeast China, and Northwest China. SEPA officials believe that establishing these local monitoring offices marks an important step forward in developing an effective environmental supervision system. Moreover, by establishing such offices throughout China, SEPA will take the helm in the oversight, management, and enforcement of environmental protection laws and regulations. This will give the central government's environmental policies greater impact at the local level, and help to harmonize oversight and enforcement efforts at the provincial and local level.

While this organizational innovation makes a big difference, SEPA had to take the equally important step of finding a staff equal to the challenge. To do so, it conducted an open examination between mid-August and early October that drew candidates from the bureau level in the national environmental protection system. From this pool, SEPA selected fifteen officials and appointed them to the eleven satellite offices, to serve as vice-directors of these offices.

These monitoring centers act on behalf of SEPA. Their major responsibilities are to supervise and inspect how national environmental policies, statutes, and standards are executed in local regions; how serious cases of environmental pollution and ecological damage are handled; how serious emergency environmental events are dealt with; how major pollution sources are monitored; how production safety and environmental protection are integrated into state-approved construction projects;[15] and how environmental law is enforced in national nature reserves (noted landscape areas and forest parks) and in national reserves for ecological function and protection.

They are also supposed to help mediate major inter-province environmental conflicts. This includes coordinating actions and resolving tensions and conflict between different regional environmental protection offices. Finally, these centers will receive and organize complaints resulting from environmental pollution and ecological destruction in land and water system that affect multiple provinces.

By projecting its power at the provincial level, SEPA's move changes the role of local officials. In fact, a "competition" between the national will and local interests started with the announcement of the SEPA plan. The first "supervision notice" regarding environmental protection issued by the Environmental Supervision Center in Southwest China described this tension as a "cat and mouse" conflict. As reported in the *People's Daily* on November 15, diesel fuel was discovered leaking from pipes at the Luzhou[16] Power Plant into the Yangtze River. SEPA categorized this event as a major environmental pollution incident, alerting the relevant departments in Sichuan Province to swiftly identify the accident's cause, those responsible for causing the accident, and summarily penalizing them in accordance with the law. It also instructed these departments to follow and monitor the inspection process and the entire manner in which administrative processes dealt with the accident and its fallout. Ma Ning, director of the Supervision Center for Environmental Protection in Southwest China, who was in charge of the on-the-spot investigation, discovered severe cases of information being withheld in connection with this event. Specifically, he found that the actual amount of diesel

[15] Translator's note: In China, such measures are typically referred to as "the three meantimes." This refers to the demand that in construction projects, supplementary facilities to ensure production safety and environmental protection must be designed, constructed, and put into operation at the same time as the primary construction project. A more detailed explication of this phrase appears in the *The China Environment Yearbook (2005)*.

[16] Translator's note: Luzhou is a city located in southeastern Sichuan Province.

fuel leaked into the river equaled 16.94 tons, or fifty times the initially reported figure of three hundred and eighty kilos.[17]

VIII. *Strengthening Protection Efforts*

Premier Wen Jiabao stated, "The environmental protection policies and the executive power backing them should be as strong as steel, rather than as soft as piece of bean curd."[18] To put it metaphorically, China is forging a sword that can cut off the black hand of environmental pollution and ecological destruction, guard our mountains, rivers, and homes, champion people's health, and facilitate sustainable development and the construction of a harmonious society. This sword is the national will and the enforcement of environmental protection. Now this sword is becoming sharper and stronger.

Various observers have noted an important change in environmental protection: the coordinated employment of a wide variety of administrative, legal, economic, and technological measures. This multi-method approach provides greater flexibility and facility in resolving problems.

A. *Policy Instruments*

After circulating the *Decision on Implementing the Rational and Scientific Analysis on Development in Order to Strengthen Environmental Protection* early in 2006, the State Council released the *Decision of the State Council on Strengthening Energy Conservation* in the middle of that year. The latter explicitly identifies the energy consumption index as an important item in assessing the performance of officials in local governments and leaders of large and medium-size firms. The *State Council's Circular on Issues Concerning the Strengthening of Land Use and Control* further clarified the responsibilities of local governments in terms of land management and arable land protection.

---

[17] Translator's note: One metric ton of oil or some other fuel = 7.33 barrels. One barrel = 42 gallons or 159 liters. In English measurements, the figures here work out to 124.17 barrels or 5,215.15 gallons. The metric total would be 19,743.06 liters.

[18] Zhou Shengxian, "Comprehensively Strengthening Environmental Policies and the Legal System, and Striving to Push Forward the Historical Transformation of Environmental Protection," *Environmental Protection* 24 (2006).

B. *Legislative Instruments*

In February, the Ministry of Supervision and SEPA jointly issued and began implementing the *Provisional Regulations on Disciplining Dereliction of Duty in Environmental Protection Activity*. The *Regulations* served as a forceful legal weapon for environmental protection by stipulating that administrative and disciplinary punishments should be meted out to those in national administrative offices and enterprises that have broken the law. Sanctions are focused on those officials who fail to perform their duties or engage in willful misconduct concerning environmental protection.

In order to intensify the strength of laws and regulations on environmental protection, and to remove restrictions on the power of law enforcement departments that confront polluting enterprises, the *Environmental Protection Law* is going to be modified fundamentally. The Environmental Protection Law is expected to become a basic national law. Preparations are being made to make or modify other laws and regulations on environmental protection such as the *Regulations on Environmental Impact Assessment for Urban Master Plans* and the *Assessment Measures for Target Responsibilities Concerning Environmental Protection*. A comprehensive and effective system of environmental laws and regulations covering all fields of environmental protection is expected to be established in five to ten years.

In order to encourage all interested parties, such as economic departments, environmental management departments, environmental NGOs, and environmental experts, to participate in the decision-making process concerning environmental protection, on February 22, SEPA issued and began implementing the *Provisional Measures on Public Participation in Environmental Impact Assessment*. The *Measures on Public Participation in Environmental Protection*, which is higher in authority and more comprehensive in the fields covered than the Provisional Measures, was being drafted during the latter part of 2006. A new decision-making mechanism for public participation provides broad space and a wide stage for environmental NGOs to play a more important role in China's environmental protection initiatives. Bringing these NGOs into the center of the environmental protection initiatives vastly expands the capacity to monitor, gather and report information, and meet other responsibilities regarding environmental oversight.

C. *Economic Instruments*

The year 2006 has seen the elevation of environmental protection from the "treatment as last resort" to the centerpiece of economic policy. This change is reflected in both the release of green GDP accounting reports and the advancement of green purchasing rules by governments.

The Eleventh Five-Year Plan allocates one hundred and forty million RMB for pollution treatment across the nation. As it stands, investment on regular treatment projects has not been fulfilled. Furthermore, investment on treatment projects in major drainage areas that have been approved by the state has not been completely fulfilled. Without these investments, pollution treatment targets cannot be achieved.

Fortunately, change seems to be afoot. In 2006, the Ministry of Finance officially included environmental protection in government budgets on ten major line items and fifty subsidiary line items. The ten major items include environmental protection management affairs, environmental supervision and monitoring, pollution prevention and treatment, natural ecological protection, natural forest protection, the "green grain" project, treatment of sandstorms and desertification, and returning pasture lands to open grasslands.[19] Under each item there are various sub-items, totaling fifty in all. This classification of expenditures, basically covering all issues related to pollution prevention and treatment and ecological protection, opened an account for environmental protection in the government's budget. This has become an important guarantee for the increase of government input into environmental protection that lays a foundation for establishing a solid fiscal base of support for environmental issues and policies.

In 2006, the central government adjusted tax policy in an effort to further bolster environmental policies. To start, the government raised the sales tax on large-engine automobiles. The government also imposed a 5 percent tax rate on disposable chopsticks and hardwood floors.

The Ministry of Finance, SEPA, and some local governments, such as that of Jiangsu, also formulated an experimental program for purchasing emission rights.

---

[19] Ma Zhong, Wang Yaoxian, and Wu Jian, "Establishment of a Fiscal System for the Environment and Enhancement of Input into Environmental Protection Is A Key to Implementing the *Decision* Made by the State Council," *China Environment News*, September 7, 2006.

To create a check on reckless and high-resource-demand economic growth, the government put in place a new pricing system on energy use. Beginning October 1, firms scheduled to be shuttered would pay an additional 0.1 yuan per kilowatt hour (kWh) during their winding-down process. Resource-hungry firms that remain in operation would face a 0.03-yuan-per-kWh hike. These prices would encourage old, inefficient firms to increase their efficiency as they wind down operations; firms that remain online would have to find ways to reduce their total demand while meeting their output goals. Incidentally, this pushes operations toward a more intensive and efficient mode of output, which fits with the long-term goals that link economic strategy to environmental policies.

D. *Law Enforcement and Inspection*

In 2006, SEPA organized and conducted a series of activities aimed at sharpening law enforcement and inspection quality. This effort involved the appointment of specific institutions to supervise the monitoring of certain major cases. By implementing this effort nationwide, monitoring brought attention and scrutiny to heretofore hidden troubles in environmentally sensitive regions. The program also established emergency-response programs in order to keep environmental accidents from becoming widespread catastrophes. Through October, a total of 1.48 million law enforcement personnel were dispatched as part of these inspections. They identified nineteen thousand enterprises in violation of environmental laws that were then penalized; another 5,036 polluters were given warnings.

To safeguard the public's environmental rights and interests, seven divisions under the State Council jointly launched an initiative to control illegal discharge of wastes. In the course of carrying out this effort, seventeen officials were held administratively responsible for negligence or failure to carry out their duties. Two were prosecuted under civil statutes.

These swift and forceful moves have made environmental protection one of the most eye-catching issues of the year.

Such efforts have come belatedly, unfortunately. Baiyangdian Lake, the largest freshwater lake in North China, boasted a water surface area of one thousand square kilometers in the 1940s, but this total has dwindled to a mere three hundred square kilometers. Apart from its shrinking area, disruption to the natural ecological functions of Baiyangdian Lake merit deep concern. In March 2006, locals reported strong odors coming from the lake, an unusually dark coloration to the water, and large numbers of dead fish in and around the lake. Inspections revealed high levels of pollutants in the lake. These pollutants had been dumped directly into the lake by nearby businesses. Because of lax monitoring, their activities had escaped notice. However, the effects were plain: more than one million USD worth of fish died and an unknown number had been contaminated. In the end, a vice district chief in the New City Zone of Baoding City, Hubei Province, was given an administrative warning; the director of the environmental protection office of this district took the blame and resigned, and two vice-directors of the office were dismissed.

E. *A United Front*

In August 2006, the State Environmental Advisory Committee and the Committee on Science and Technology, under the aegis of SEPA, began operating. Through these organizations China set up its most extensive "united front" for environmental protection. These two "committees" consist of forty-nine Chinese environmental protection experts and thirty-seven environmental technology experts; thirty of these individuals are academicians affiliated with the Chinese Academy of Sciences and the Chinese Academy of Engineering. Qu Geping, the first Minister of SEPA and the President of China's Environmental Protection Foundation, is also a member of the State Environmental Advisory Committee. He pointed out that

> The most serious mistake we made in dealing with previous problems lies in the lack of rational and democratic decision-making. Now, the public sense of democratic decision-making has been considerably enhanced. Controlling environmental pollution is the main task of environmental protection organizations. Clearly, this is a formidable challenge. It is far from enough to rely solely on the strength of SEPA; rather, we need participation of the entire society. In this sense, the founding of the two "committees" will definitely produce a significant impact on China's environmental protection.

IX. *Striving for Change*

Perhaps the tension between development and protection will never end. While "development is of overriding importance," development without sound and rational analysis does not generate true development. Only development grounded in science and rational analysis is of overriding importance. More and more Chinese have come to realize that challenges and tensions involving environmental conditions are interrelated, and therefore require systemic treatment. This includes the need to deal with resource bottlenecks that slow production and raise production costs, as well as more obvious problems caused by environmental pollution and ecological degradation. Engaging these problems within a comprehensive framework is essential in order to realize the goal of healthy, sustainable development, in both the economic and social sense. If a breakthrough is to be made in our work on environmental protection, governments at various levels must abandon their old mode of achieving high economic growth through vast resource input and vast pollution output. They must discard the erroneous concept of regarding economic growth as their first, and sometimes only, priority, which drives them to pursue GDP growth blindly. Rather, they should aim to achieve sustainable development within the framework of resource conservation and environmental protection.

In recent years, we have heard many voices call for the strengthening of environmental protection in China. In 2006, such voices became louder and stronger than ever before. Quasi-governmental organizations like the World Bank and the China Council for International Cooperation on Environment and Development (CCICED), as well as various non-governmental organizations, advised China on setting up a ministry of environmental protection. Professor Gu Haibing, from Renmin University,[20] and other experts and scholars both at home and abroad have echoed this call by proposing the establishment of an environmental protection "super agency" made up of SEPA and other relevant departments. The report, called *Research on Administrative Capability Concerning the Chinese Environment*, which CCICED recently completed, states that, considering China's territorial size, population, and environmental problems, its national environmental protection system is relatively weak, especially in terms of human resources. Many experts

[20] Translator's note: Also known as "People's University."

also pointed out that in light of China's difficult environmental condition and its future development prospects, it is imperative to emphasize and implement environmental treatment built around "firm governmental intervention and extensive social participation" by strengthening human and other resources.[21]

In 2006, the nation will challenge the extensive economic growth model. In doing so, China will continue to advance its environmental protection efforts even as they encounter obstacles and difficulties. Part of this certainly hinges on making changes in government structure, in particular by raising SEPA to the ministerial level and enlarging its staff. In the final analysis, though, all environmental protection forces must successfully navigate the transitional period at hand and work together more effectively.

[21] Xia Guang, "Thoughts on Establishing Regional Supervision Centers for Environmental Protection," *China Environment News*, August 28, 2006.

PART ONE

THE ENVIRONMENT AND SOCIETY

Two steps have had a broad and deep effect on China's environmental protection efforts in 2006: new conceptual thinking on the value and motivations for extending environmentalism, and the introduction of the Green GDP. These two broad-brushstroke contributions have led to more concrete and refined changes: more systemically-organized work in the field, and more reliable and consistent fiscal support stand out as two of the most important green initiatives included in The Eleventh Five-Year Plan. A group of new legal measures has also given greater form and substance to policy in this area: *The Law on Renewable Energy* and *The Provisional Regulation on Public Participation in Environmental Impact Assessment* stand out as the most important environmental laws introduced during the year.

Noteworthy as well are the changes taking place in rural areas. The policy of "building a new socialist countryside" has aroused public concern largely because of this effort's environmental impacts, but also because of the social, cultural, and economic dislocation that comes with it. However, numerous success stories can be found and may light the path to developing even more economically and culturally vital countryside communities that balance these human considerations with ecological conditions and demands.

Environmentalism has worked its way into other parts of Chinese society, sometimes provoking conflicts and other problems. For instance, deep shifts in values and ethics have improved animal rights, but at the same time, wild fauna continue to live under duress in many cases. On the other hand, positive change came about in the form of "Green Filming," which grew out of a vociferous public reaction to environmental abuses caused by the film crew working on *The Promise* near Bigu Lake in Yunnan Province's aptly named Shangri-La County.

CHAPTER TWO

CHALLENGES OF AND PROSPECTS FOR GREEN GDP ACCOUNTING

Zhang Ying

Rapid economic growth in China has been accompanied by ever-increasing consumption, waste, and ecological destruction. In essence, these costs have not been integrated into economic measurement. Calculating the Green GDP has been proposed as a way of making the link between economic growth and the array of costs generated by growth. Implementing green GDP accounting systems would have numerous, important economic, political, and social implications. Further complicating matters, its calculation and implementation pose immense difficulties and challenges, which require proper understanding and concrete action on the part of nearly all Chinese, from government ministers, policymakers, and bureaucrats, to businesspersons, entrepreneurs, and even average citizens.

Key words: Green GDP, national economic accounting system, science-based development

I. *Green GDP in 2004*

On September 7, 2006, the State Environmental Protection Administration of China (SEPA) and the National Bureau of Statistics of China (NBS) jointly issued the *Green National Accounting Study and Report (2004)*. According to this report, economic losses caused by environmental pollution amounted to 511.8 billion RMB, or 3.05 percent of 2004 GDP. Losses caused by water pollution amounted to 286.28 billion RMB, or 55.90 percent of total pollution-generated losses. Air pollution losses amounted to 219.8 billion RMB, or 42.90 percent of total losses. Losses from solid waste and pollution accidents came to 5.74 billion RMB, or 1.20 percent of the total.[1] Figure 2.1 summarizes these findings.

The concept of "green accounting" began in 1993. It started as a way of incorporating the value of natural resources and the environment into the national accounting system. By developing these more comprehensive

---

[1] Gu Ruizhen and Wang Yani, "Economic Losses Caused by Environmental Pollution Account for 3.05% of GDP in 2004," September 8, 2006, http://www.xinhua.net.

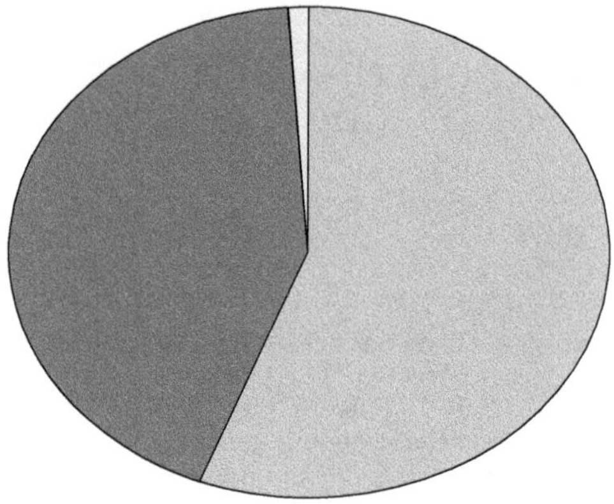

Figure 2.1: Shares of economic losses caused by environmental pollution in China in 2004

statistical indexes, market prices truly reflect the value, as well as the costs, of economic activity. Unlike standard economic accounting, which often values certain environmental interactions as "external"[2] to the market, green accounting integrates these elements into a more complete economic model.

The Green GDP report was released at a time when China had committed herself to the pursuit of development based on scientific principles and rational decision-making[3] within the structure of the so-called "cradle-to-cradle economy."[4] The report describes the grave environmental situation in China in detail, and links this condition to the

[2] Translator's note: Typically, "byproducts" of economic activities not included in the market system are called "externalities."

[3] Translator's note: Chinese media often refer to rational economic decision making as "scientific development." The idea is that policy should be based on current scientific knowledge and a market-based decision, production, and allocation system that assumes rationality.

[4] Translator's note: Chinese media typically refer to this notion as the "cyclical economy." Since this phrase is not used in the West, the term "cradle-to-cradle economy" will be used throughout this essay. This term was coined by the designer Bill McDonough to describe design and economic processes that regenerate themselves, that ideally generate zero waste, and that integrate recycling as part of the productive process from the design stage forward. See *The China Environment Yearbook (2005): Crisis and Breakthrough of China's Environment* (Leiden: Brill, 2007) for a more thoroughgoing explanation of this term and its use.

economic growth system in place. The report raises troubling questions regarding the continued use of current economic growth policies.

China suffers from increasingly severe environmental problems. From 2001 to 2005, the volume of wastewater discharged increased from 43.3 billion tons to 52.3 billion tons, carbon dioxide (CO_2) emissions rose from 19.478 million tons to 25.49 million tons, and soot emissions climbed from 10.698 million tons to 11.82 million tons.[5] Statistics show that although China's GDP is only 5 percent of the world's total, it consumes 20 percent of the world's aluminum and copper, 30 percent of its steel, iron ore, and coal, and 45 percent of global cement output. From 1992 to 2004, the volume of China's carbon dioxide emissions rose more than 33 percent, making China the world's second-largest carbon dioxide producer. CO_2 output continues to grow, rapidly, and in a few years China may be the top emitter of this gas. Chinese cities rank notoriously high among the most polluted cities worldwide. About 20 percent of China's population lives in severely polluted areas, and 70 percent of all rivers and lakes are contaminated.[6] This leads many experts to believe that calculating estimated losses at 3.05 percent of GDP falls well short of total economic, environmental, and social losses.

Green GDP accounting is an important component of green accounting. It includes both resource depletion costs (affecting resources such as soil, minerals, water, forests, grasslands, and fisheries) and costs arising from environmental degradation (costs produced by environmental pollution and ecological damage) to determine the impact of economic development on the environment. Lack of funding and organization limited the 2004 accounting results to include only the costs of environmental pollution (primarily water and air pollution). The results based on these data provide only a partial adjustment to GDP; this is not the Green GDP in its full sense.

[5] The Chinese National Bureau of Statistics, *Chinese Statistics Excerpts, 2006* (China Statistics Press, 2006), 202–203.

[6] Brian Bremner, "The Influence of China," *Reference News* 8 (October 22, 2006).

II. *Difficulties Encountered in Developing Green GDP Systems*

Green GDP accounting overcomes limitations in traditional GDP accounting by revealing natural resource and environmental costs incurred through economic activity. This more complete body of data makes green GDP accounting a potent tool in planning economic and social development policies and systems. This in turn leads to a more comprehensive and truthful evaluation of socioeconomic policies, in particular those centered on growth and sustainability.

Two decades of reform and liberalization have moved China onto the economic fast track. GDP has tripled during this period; China now trails only the United States, Japan, and Germany in GDP rankings. The flip side of this coin, however, is not so bright and shiny. Growth has been achieved while depleting natural resources and committing widespread environmental damage. Energy consumption per unit of GDP in China exceeds the global average (excluding China) by more than 400 percent. Mineral input per unit of output ranges from 200 to 300 percent of the rates observed in developed countries like the United States and Japan.

Environmental damage through pollution stands at unprecedented levels. In 2005, wastewater discharge exceeded environmental carrying capacity by about 27.59 billion tons. From 2001 to 2005, carbon dioxide and soot emissions climbed by 30.87 percent. Chemical oxygen demand (COD) increased by 10.49 percent and ammonia and nitrogen emissions grew by 0.58 percent and 19.81 percent, respectively. Sulfur dioxide emissions exceeded environmental carrying capacity by about 200 percent.[7] Carbon dioxide emissions per unit of GDP in China were twice the ratio calculated for Japan.

China's huge population exacerbates these problems because the general standard of living continues to lag behind that of more developed countries. Per capita GDP barely exceeded one thousand USD in 2003. Even after five years of robust growth, per capita income was 13,985 RMB (about one thousand eight hundred USD) in 2005. In comparison, output per capita for the entire world had already exceeded 5,199 USD by the year 2000.

China continues to follow the traditional model of industrialization characterized by high consumption, low efficiency, and high pollution

[7] The Chinese National Bureau of Statistics, *Chinese Statistics Excerpts, 2006* (China Statistics Press, 2006): 202–203.

as practiced in the West in the last two hundred years. This has produced enormous challenges in meeting natural resource demands and recognizing resource constraints. These tensions shape environmental protection and population control policies. Living conditions for Chinese citizens continue to improve economically, but in many ways the quality of life has been degraded. To correct these problems and to resolve these tensions, China needs to adopt development principles and policies based on scientific insight and rational decision-making. Moving from traditional GDP accounting to green GDP accounting would be an enormous step forward in accomplishing these things. By balancing measures of economic activity against this activity's impact on the environment, and by considering the quality of economic activity and not just its quantity, resource problems, population pressures, and quality-of-life issues can be addressed in one stroke. In short, the environment and individual and social interests should not be sacrificed on the altar of Mammon.

In the System of Integrated Environmental and Economic Accounting (SEEA), the United Nations (UN) points out that green GDP accounting is meant to measure the impact of the economy on the environment and to reflect natural resource and environmental conditions. However, the deeper impact of this accounting is to alter national accounting procedures so that economic planning, behavior, and decisions change. Individuals must come to realize that economic growth cannot be sustained at the expense of the environment, since sustainable economic and social development can occur only while maintaining environmental health and balance.

Green GDP accounting guarantees sustainable development by emphasizing the integration of various economic, environmental, and social considerations, and the data that measure them. These indicators allow national-level policymakers to begin the work of internalizing environmental costs within the market structure and decision-making processes.[8] Potentially, this can transform individual, group, and organizational behavior.

[8] Translator's note: A variety of resource and environmental costs are not fully integrated into the price-driven market system. Because they typically exist outside of this system, economists call them "externalities." By moving these processes and goods inside of market mechanisms, prices more fully reflect value, and rational decision making makes use of this richer body of information to help decision makers, from ordinary consumers to macroeconomic policy strategists, to make more useful and responsible decisions.

Since the 1960s, when the "environmental revolution" began, experts on natural resources and environmental economics have explored ways to internalize the external costs in order to achieve a more efficient, fair, and sustainable utilization of resources. Green GDP accounting supports and complements such efforts.

Green GDP accounting helps to forcefully present fresh views of development and growth. It is built on an overlapping foundation of principles that mutually reinforce one another. First, this system of accounting respects and integrates the best scientific knowledge currently available. Second, it applies a human-centered ethos in place of the currently-applied economic model that centers almost exclusively on growth. Third, its systems are integrated and integrative; the economy is placed within the context of a broader environment, rather than subjugating the environment to serve economic ends. Because of this, green GDP accounting is more comprehensive and complete. This allows policymakers, business leaders, bureaucrats, and average citizens to operate and interact in a more balanced and coordinated fashion.

Green GDP accounting unites economic growth and environmental protection. A truer, clearer, and more complete picture of economic output and activity is balanced against the costs of achieving these things, which include resource depletion and ecological impacts. Replacing the traditional GDP measurement with the Green GDP Index helps to disentangle confusion over "development" and "growth."

Implementing green GDP accounting will have a direct impact on government operations. It will hold policymakers, regulators, bureaucrats, and agencies involved in environmental oversight and enforcement accountable to citizens and the stated mission of their organizations. By measuring GDP in greater depth, we can better assess the performance of government officials. Furthermore, we can better assess whether the policies and activities they promote truly serve the mission of the state and the country.

This change in evaluating government officials' performance will affect departments and officials at various agencies and government levels. Because of this, implementing the green GDP system has encountered some resistance. This is the case even though SEPA and other related governmental departments have worked hard to lay the groundwork for employing the change. For instance, they have educated officials on the relationship between green GDP accounting and the political futures of administrators and bureaucrats. However, technical problems in implementation and inadequate awareness of the benefits of the system caused major setbacks. Many officials worried that the new criteria would be detrimental to local and regional economies.

After all this effort, the process has been brought to a halt. Since the process stopped moving forward, no consensus has emerged that would make the Green GDP the official measurement of output and economic activity, and therefore of government officials' performance. The questions raised and the problems encountered underscore the fact that adopting a new policy system will create doubt and uncertainty in many government quarters. It would seem that supporters of this change must produce more research regarding the impact of the change, and provide a clearer picture of likely outcomes to help allay the fears of those who oppose it.

In 2005, SEPA and the NBS launched a green GDP accounting pilot program in ten provinces and municipalities, including Beijing, Tianjin, and Hebei. On December 9, 2006, the technical team that released China's first Green GDP Report won a special award called "Leading Green Figures in China in 2006." The judging committee commented, "Though the first step is far from perfect, it is nevertheless one major step towards building a development system grounded in scientific knowledge and rational decision making." When receiving the award, the representative of the technical team said that green GDP accounting was not supported by all Chinese provinces, and that, in fact, some provinces and cities even asked to withdraw from the pilot program. However, Mr. Pan Yue, vice minister of SEPA, remarked, "Even if there is only one province left in the pilot program, the Administration would stick with it and see it through to the end."[9] The technical team will release China's green GDP accounting results from 2004 in the first half of 2007. This time lag demonstrates how difficult and complicated the work is.

III. *Problems Facing Green GDP Accounting*

Producing a Green GDP Report is a complicated undertaking. The process used to create the report involves diverse fields of study and research such as economics, statistics, accounting, resource management,

[9] Ma Li, "The Green GDP Technical Team: Some Provinces Exit the Pilot Program for Green Accounting," *Beijing News*, December 12, 2006.

ecology, and environmentalism. After more than a decade devoted to researching green accounting, the UN formulated the SEEA. Other international organizations have conducted similar research. In spite of these efforts, no standard on the matter has been widely adopted.

A. *Shortcomings in the Framework*

At present, research on and practices of green accounting have been carried out at state and local levels using the SEEA as a guide. As noted above, this is the current green GDP accounting standard used by the UN and a number of other organizations. However, disagreements among researchers have led to the development of distinctly different accounting methodologies. In particular, relationships between different accounts, subjects to be counted, and means for determining valuation vary markedly. These conflicts have derailed widespread adoption. To correct these problems, achieve a usable standard, and promote adoption of these systems, the NBS and relevant government agencies need to formulate a comprehensive framework and implementation system for green accounting. This will serve as a much-needed guide to refining and harmonizing existing green GDP accounting practices. For example, SEPA and the NBS formulated *The Framework for Integrating Environmental and Economic Accounting in China*[10] in 2006; the State Forestry Administration and NBS formulated *The Framework and Content of Forestry Accounting within Green GDP Accounting.*[11] These two documents share common ground as well as disparities. Some localities, such as Beijing and Daxing'anling, in northeast China, formulated their own research framework and green accounting systems based on independent research projects. Because of these disparities, research results cannot be compared, since research subjects and methods vary from one study to the next.

[10] The State Environmental Protection Administration (SEPA) and the National Bureau of Statistics, *Research Report on Green GDP Accounting in China* (July 2006), 56–118.

[11] The State Forestry Administration and the National Bureau of Statistics, *The Framework and Content of Forestry Accounting within Green GDP Accounting* (September 2006).

B. *A Simplified Understanding of Green GDP*

According to the SEEA, mankind's activity has had a profound impact on the environment in the twentieth century. Concerns regarding the impact of each country's economic activity on the environment continue to mount. Additionally, it has become clear that maintaining economic growth and improving human welfare depend on the benefits provided by the environment. Green accounting expresses this connection explicitly in quantitative terms. This body of data can help society move one step closer to finding real, lasting solutions to many environmental problems. Therefore, many people believe that environmental problems can be solved by implementing a meaningful form of green GDP accounting. At the very least, green accounting can help reduce the costs and problems associated with resource depletion and environmental degradation.

In fact, green GDP accounting is an integrated system that goes beyond simple adjustments to gross economic measures. It not only involves adjustments to macroeconomic indicators, such as GDP and Net Domestic Product (NDP),[12] by counting resource and environmental factors, it also accounts for assets, liabilities, national wealth, and net savings within an ecological context.

If our knowledge of green GDP accounting is superficial, green GDP calculations will inevitably become a numbers game. To meet various traditional goals, some localities may seek quick successes by "playing" the system, in order to achieve a strong green GDP result. To satisfy this demand we can expect to see green GDP accounting "specialists" appear in various markets. This will likely produce serious mistakes in applying the system, and the policy may lose its rigorousness. To avoid this outcome, the system must be thoroughly understood and correctly applied. Those seeking to manipulate the system from the government side or the accounting side must be weeded out of the system and sanctioned if necessary.

Taking these steps will help China put relevant scientific findings to use when forming policies. Performance of government officials and administrators can be clearly and fairly assessed. However, the failure to do this leaves the country and its localities vulnerable to ongoing miscalculations, poorly formed policies, and unscrupulous business and government leaders.

---

[12] Translator's note: NDP equals GDP less asset depreciation.

C. *Redundant, Shallow Research*

Research on green GDP accounting started early in China. However, many researchers rehashed approaches and recycled results. Few undertook in-depth studies. For example, three groups at the state level have studied green accounting in the context of forestry resources since 1987. These studies have been conducted "independently" at different times. And this does not count other studies on green forestry resource accounting conducted by local agencies and various other research institutes. This redundancy wastes human, financial, and material resources. Rather than breaking new ground with each study, studies only mirror available results and methods in a festival of mutual admiration and groupthink.

Other sectors have seen a similar pattern. Literature reviews of green accounting research in China reveal the scarcity of innovative and profound work in the field. Specialists in green accounting worthy of the name are few and far between. The field desperately needs to recruit committed, talented, and pragmatic persons who take a no-nonsense approach to doing research and to seeing that the results of research have a measurable impact on policymakers, business leaders, and public discourse.

D. *Underdeveloped Accounting Methodologies Used in Green GDP Calculations*

The methodology of green GDP accounting is very complex and still immature. Many key methods are still being researched and refined. Key issues, such as pricing environmental resources, have yet to be meaningfully codified in the system.

Technically speaking, GDP accounting should be based on market transactions. Prices form the lynchpin of this analysis. Various inputs and processes interact through the market to determine these prices. However, problems within the pricing process call stated prices into question. Note that in some cases, prices do not capture the full value—or costs—of many inputs and outputs. Generally speaking, environmental inputs or outputs affecting parts of the environment[13] that are not clearly defined by property rights tend to have prices that

[13] Translator's note: Typically, markets undervalue these elements and their interactions. At worst, markets assign a value of zero. Even though this is an extreme case, it is also a common outcome.

diverge from their true value.[14] For example, the value of flora and fauna in a forest, the ecosystem that binds them together, and the various environmental functions produced through these interactions cannot be directly or precisely evaluated through the market. Therefore, virtually any price for these elements and their interactions will be fundamentally miscalculated.

In spite of its fundamental importance to life, water resources are undervalued. Because of this, water pollution is rampant. And while creating the problem is easy and "cheap," correcting it is costly and far more explicitly stated. Specifically, restoring China's lakes and rivers to relatively good Level II quality will require tens of billions of RMB. Note, however, that it is not clear that the water resources themselves would enjoy equal value *ex ante*. Measuring this divergence is a subject of continuing research.

Although environmental economists have improved their theories and methods of resource evaluation, resource values still diverge from their market prices. Resolving this divergence remains one of the central problems in developing, much less implementing, a meaningful green GDP accounting system.

IV. *Trends in Green GDP Accounting*

A. *Establishing a System of Sustainable Development is a Long-Term Process*

Green GDP accounting is of enormous significance in revealing resource depletion and environmental degradation caused by economic growth, reflecting the impact of social and economic development on natural resources and environments, and coordinating and balancing relations between humans and nature. It is useful for establishing a viable system of sustainable development.

The coordinated and balanced development of the economy, society, and environment is a complicated systematic project. So is the building of a comprehensive, coordinated, sustainable, humane, and rationally-organized economic system. By itself, green GDP accounting cannot

[14] Translator's note: Some economists, particularly theorists, will balk at this notion. Within a complete, fully developed market, prices and value are identical. Clearly, the argument here rejects this notion.

solve these various and interrelated problems. We also need institutional, legal, regulatory, and policy tools, especially from the public sector. From the private sector, more productively- and reasonably-applied capital input, management decisions, human resources, scientific and technological findings, and insight will contribute significantly to resolving the issues China faces. We need to devise systematic theories as guidance, establish a sustained and well-coordinated, science-based outlook on development, and then put these theories and policies into practice.

Green GDP accounting captures short-term trends as well as long-term processes. It provides a multifaceted, quantitative description of economic, societal and environmental development in a certain period. Because it is comprehensive and systemic, it typically provides information, and therefore acts as a foundation for policy and decision making. Yet for all its analytical benefits, green GDP accounting must be backed up by meaningful action regarding economic choices and behavior if we are to solve environmental problems in China.

B. *Large-scale, Comprehensive, and Basic Research Needs to be Conducted*

Since its introduction, the green GDP concept has been developed and improved gradually. The basic framework and formulation have become well articulated, but many finer details remain sketchy. Regulations have not been clearly described or integrated into the broad framework, policies have not been thoroughly organized, and assessment and evaluation standards remain in question. To address these shortcomings, policymakers and administrators need to systemize and standardize green GDP accounting. Policy suggestions on this point are described immediately below.

1. *Developing and Improving Related Laws and Regulations* Even though the number of environmental laws and regulations continues to increase, it is not clear how effectively these support or contribute to the development and implementation of green GDP accounting. Various elements have yet to be addressed: defining specific accounting methods; clearly establishing the scope of green GDP accounting; precisely classifying environmental costs; providing standards of accounting deductions; accurately calculating costs and prices; and developing a system for spurring the adoption of green GDP accounting. In short, these tools need to be included in an environmental accounting and environmental auditing system. As noted earlier, once this

system is in place and widely applied, it must be used to produce real changes in behavior. Among other things, this means integrating the Green GDP Index into a system for judging government officials' effectiveness. This is one of the most important auditing functions of the green GDP toolkit.

2. *Improving Related Data Collection and Organization* At present, resource oversight, monitoring departments, and departments that gather and analyze statistics do not effectively coordinate their actions and systems. To correct this problem, we should establish a coordination mechanism for collecting, analyzing, and sharing statistics needed to calculate the Green GDP and to implement policies based on these calculations. A wide array of data needs to be gathered and sensibly collated: information on environmental quality and ecological interactions; water, energy, and land use and conditions; calculations of mineral reserves and use; descriptions of forest conditions and trends; and agricultural measures of output, quality, and prices. This will go a long way toward improving green GDP accounting standards and measures. It will help to fashion meaningful and credible plans based on green GDP measures. All of this should promote implementation of the green GDP system.

3. *Improving Evaluation Standards and Completing Foundational Work* Since green GDP accounting is complex and comprehensive, we must take special care in establishing its base. This means that identifying best practices, and standardizing systems based on these practices, must precede much of the implementation work. Once this is accomplished, formulating and improving the ways that these tools can be used in program and worker evaluation can proceed apace.

C. *Development Trends of Green GDP Accounting*

Considering practices at home and abroad, we see the following trends developing in green GDP accounting:

1. *Improving the Entire System of Data Collection, Collation, and Analysis* Green GDP accounting is a part of the SEEA, which is a comprehensive and internally consistent accounting system. Concepts within the system are harmonized and balanced. Methods and definitions for classifying accounts are mutually reinforcing. Economic information used within

the SEEA aims to be compatible with information provided by the System of National Accounts (SNA). Using the same information from both systems will increase the value of environmental and economic information. Using the same information in different ways will facilitate analysis and utilization of information in an analytically consistent framework. Therefore, environmental statistics need to be collected and provided in a form useful for the SEEA. This includes information on industrial classifications, which may or may not be consistently collected and/or defined across different agencies. Therefore, environmental data used for the SEEA can be integrated into other policymaking, research and analysis, and decision-making and implementation functions.

2. *Aiding in Policymaking and Decision Making* Green GDP accounting should lead to better policy and decision making. As stated earlier, green GDP accounting and the SEEA are designed to promote sustainable economic, societal, and environmental development. They do so by presenting a comprehensive and systemic body of data, and by drawing relationships between economic activity, societal needs, and environmental conditions. By providing a clearer, more complete, and more richly-integrated and -interrelated view of these conditions, policy formulation and decision making can improve.

3. *Integrating Human and Social Capital into the Analysis* Wealth refers to capital stocks. Typically, economists and accountants focus on productive capital, such as industrial assets, in their analyses and calculations. However, human, social, and natural capital contribute mightily to total wealth. Sustainable development works to maintain, and where possible increase, per capita wealth by protecting sources of wealth as well as promoting processes for generating wealth. Even the SEEA emphasizes production and natural capital while deemphasizing the importance of human and social capital. This oversight cries out for further research; such work could be of major consequence. At the very least, such research should examine the causes and consequences of resource depletion of these two particular forms of wealth.

4. *Moving Toward Standardization and Systemization in Green GDP Accounting* In spite of the numerous difficulties and challenges it may face, green accounting, as a basic source of information and analysis

related to sustainable development policy and decision making, will become more standardized and systemized. Since this tool, in some form or another, is likely to be more widely implemented, we should prepare to integrate its methods and techniques into a broad range of data collection methods and analytical practices.

APPENDIX 1

LINK: THE HISTORY OF GREEN GDP ACCOUNTING

Research on green accounting in China dates back to the 1980s, when Mr. Li Jinchang, former deputy representative of China to the United Nations Environment Program and director of the Department of Technological and Economic Research at the State Council's Development Research Center, pioneered research in the field. He chaired the research on natural resource accounting and its integration into national economic accounting in China. This included accounting research regarding forest, mineral, grassland, soil, and water resources.[15] In particular, the forestry accounting effort led by Mr. Li from 1989 to 1992 was designated by the Ministry of Forestry as a key research project and received financial support from the World Resources Institute (WRI) and the Ford Foundation. Through the efforts of Kong Fanrong, He Naihui and others, the project yielded notable results that were included in the *Handbook of National Accounting* compiled by the United Nations Statistics Division. As such, this sample case was distributed to and studied by other countries.

In 1994, the State Science and Technology Commission organized scholars and experts from the Ministry of Geology and Mineral Resources, the Ministry of Forestry, the State Education Commission, the Chinese Academy of Sciences, the State Oceanic Administration, the State Land Administration Bureau, and the Ministry of Water Resources to conduct research on natural resource accounting on a large scale. This effort included an accounting of mineral, water, marine, land, forest, wildlife, grassland, and recycling resources. After carrying out complementary preliminary work aimed at firming up the definitions of prices, completing an examination of physical accounting, refining an accounting of value and prices, and fusing this work with studies of accounting theory, methods, and techniques, the group generated an in-house report, the *Study of Chinese Resources Accounting*.

---

[15] Li Jinchang, Jiang Wenlai, Jin Leshan and Ren Yong. *Study on Ecological Value* (Chongqing: Chongqing University Press: 1999), 1–4.

Since 2000, with the financial support of the International Tropical Timber Organization, Mr. Hou Yuanzhao, a research fellow from the Chinese Academy of Forestry, conducted research on value accounting concerning tropical forest resources and then integrated findings from this work into national economic accounting systems in China. His publications include *Valuation and Accounting of the Forest Environment* and *Forest Resource Accounting*. His research findings have been translated into English, French, and Spanish. Forest resource accounting, based on the broader project of developing green accounting methods and systems in China, has far-reaching international influence.

However, the research cited above focuses primarily on the accounting of natural resources. Since the 1990s, the environmental crisis in China has been worsening. The accelerating growth of urbanization and industrialization shows no signs of abating. The mode of economic growth used to drive and supply these processes consumes vast energy and resource inputs, while generating huge volumes of pollution and waste. Taken together, these things aggravate environmental problems. Experts and scholars have begun a desperate search for solutions needed to meet these environmental challenges. To bolster their efforts, the Chinese government has put green accounting on the policy agenda. The Green GDP Index will be a key index in assessing officials' performance.

The NBS attaches great importance to research on green GDP accounting. Their efforts have taken many forms. For instance, in 2001, the bureau conducted a pilot program in Chongqing to account for the damage caused by water pollution and industrial pollution. In 2004, the NBS and SEPA established a panel on research into green GDP accounting. In the same year, the NBS, the Ministry of Forestry, the Chinese Academy of Forestry, and the Beijing Forestry University officially launched research to bring accounting of forest resources in line with the green GDP accounting framework in China.

APPENDIX 2

RELATED COMMENTS: SHOULD WE EXPECT GREEN GDP TO BE WIDELY IMPLEMENTED?

Gao Yuan

So far, no country has calculated the Green GDP comprehensively based on total resource consumption and environmental costs alongside economic activity. Major difficulties in environmental accounting regarding the means for effectively quantifying the value of environmental resources and service functions of the environment continue to thwart such efforts.

Behind the scenes, though, changes affecting green accounting continue to surge forward. Although many people still attach paramount importance to economic growth rates, a growing number of thinkers now see the limitations in relying solely on this measure as a sign of economic and social well being. This has prompted more discussion on whether a new evaluation standard should be established, and what it should be. More voices are calling for the establishment of an alternative standard reflecting the interaction between the economy and environment in a way that traditional GDP measures overlook.

Interestingly, when Zhejiang's GDP exceeded one trillion RMB, its statistics department issued a report entitled *A Cost Analysis of GDP Growth in Zhejiang Province*. This report described the province's reliance on extensive growth and that model's emphasis on high input, high consumption, and high emission economic activity.[16]

The discussion of new GDP measures has become more visible; many people have adopted a more thoughtful, sober, and objective

[16] Translator's note: The "extensive" growth model is predicated on rapid growth. Typically this approach requires large inputs to achieve this growth. Operations often work inefficiently. Worse, this production model tends to produce large amounts of pollution and wastes. This economic approach stands in contrast to the "intensive" form of growth called for by many scholars, environmentalists, and increasingly, policymakers and business persons. The intensive form of growth seeks to achieve GDP growth by moving up the value chain, improving the efficiency of operations, and by moving resources into "knowledge and information" industries.

attitude towards traditional GDP and GDP growth measures. These persons have come to realize that overemphasis on GDP growth, and equating development with economic growth, have produced many economic and social problems, in particular the growing income and wealth disparity, particularly between rural and urban areas. These are monumentally complex problems, and solving them will be tremendously difficult. The question remains, however, whether green GDP accounting will produce substantive improvements in economic, social, and environmental problems, or whether we have naïvely put our faith in a "silver bullet" approach that will ultimately fall short.

Theoretically speaking, sustainable development denotes the harmonized development of the economy, the environment, and society. Green GDP simply adds environmental considerations to GDP, and reflects the interaction between the economy and the environment to some extent. However, we should take note that the Green GDP seems to overlook the interactions between the economy and society, and between the environment and society.

From this perspective, we should not overestimate the role that the Green GDP might play in framing and bolstering sustainable development. This in no way discounts its potential importance, however. As with traditional GDP measures, their limitations should be remembered while holding their importance firmly in mind.

Simply put, it is unreasonable to demand that green GDP accounting serve as a panacea. No accounting system can do this. The key limitation is that it can hardly quantify the quality of public services provided by the government, which is the core of this policy. Therefore, when used as a criterion in evaluating government officials' performance, it has clear shortcomings, even as it can provide some insight into guiding government operations.

In spite of these limitations, calling for green GDP accounting is a step in the right direction. The fact that growing numbers of people do so offers encouragement that deeper changes may be afoot. At the same time, we must remain thoughtful and critical. We should examine what it can and cannot achieve; we must learn how to use it properly and appropriately. The Green GDP is expected to become a new link in and momentum for transforming the functions of the government and establishing a more effective environmental supervision system. We must use it well to achieve such lofty outcomes.

Source: *People's Daily*, May 30, 2006

CHAPTER THREE

PAN YUE'S REFLECTIONS ON THE ENVIRONMENT

Hu Kanping

China has begun implementing a rational, science-based environmental policy[1] that seeks to support efforts to build and strengthen societal harmony and balance[2] while enriching environmental consciousness and academic research. In 2006, two of Pan Yue's publications, "The Harmonious and Environmentally Conscious Society"[3] and "On Socialist Ecological Civil Society,"[4] received broad attention from different social circles as the latest and most forceful theoretical achievements in the field.

Key words: harmonious society, environmentally conscious society, ecological civil society, socialism

I. *Pan Yue's Environmental Theory*

Pan Yue, vice minister of the State Environmental Protection Administration (SEPA), was nominated by many to be one of the leading Green Figures in China for 2006 in an online contest. Unfortunately,

[1] Translator's note: In many translations, the phrase "scientific outlook" is used to mean decision and policy making based on scientific principles and rational information processing and decision making. However, since this phrase is largely unknown in Western writings, an alternate phrasing is used that more clearly communicates the underlying idea.

[2] Translator's note: The phrase "harmonious society" is a commonly-used term in Chinese writings that means far more than social cooperation and calm. The words "harmonious society" imply and suggest a wide range of related actions and conditions: building cooperation and coordination among individuals and organizations; successfully resolving social tensions and seeking balance between various social and economic players; and avoiding and suppressing conflict while engendering peacefulness and level-headedness. This phrase suggests a more fluid kind of egalitarianism; equality is not a static state, but rather a process. The Chinese mean these things in a comprehensive way: such values should be at work within the individual, families, the society, and even the international sphere. This rhetoric draws upon Confucian values, and to some degree, reframes the dialectical values of doctrinaire communism.

[3] Translator's note: This phrase is sometimes translated as "environment-friendly society."

[4] Pan Yue, "The Harmonious and Environmentally Conscious Society," *China Economic Times*, July 4, 2006. "On Socialist Ecological Civil Society," *China Economic Times*, September 26, 2006 and September 28, 2006.

the sponsor of the poll rejected his nomination because he is a "ministerial-level civil servant," which did not meet the requirement on candidates. Some citizens expressed doubts regarding his exclusion from the list: for them, the Green Figures in China for 2006 should be those individuals "whose thought, speech, actions and decisions have a significant influence in environmental protection," according to contest requirements. Pan Yue's contributions and achievements in this field fulfill these aims. In particular, his theoretical articles on environmental and social development are especially powerful in their impact on the field.

Some attribute the impact of his thinking to his privileged position as vice minister of SEPA. This is a specious and simplistic conclusion. His theories stand on their own, regardless of his position. Moreover, his theoretical insights are built on his practices at SEPA; his work at SEPA puts his theoretical notions into practice. He is both a thinker and a practitioner. From his work experience, Pan has drawn inspiration for his deeper thoughts; from his extensive reading of Chinese and foreign theoretical and academic works, he has extracted the essence of their messages and applied them in the real world.

This interplay between the theoretical and the applied has driven Pan to be a prolific writer. In recent years, he has published "A Call for Chinese Firms to Take Responsibility in Becoming More Green," "Environmental Protection and Social Equity," "Environmental Protection and Public Participation," "On Environmental Protection in Science and Technology," "Environmental Protection Targets and Officials' Performance Evaluations," "On the Green GDP," "Sustainable Development and the Transformation of Chinese Civilization," "Environmental Culture and the Revival of the Nation," "Strategic Environmental Impact Assessments Must Be Incorporated into Macroeconomic Decision-making," "Some Thoughts on Building A Harmonious Society," "The Harmonious and Environmentally Conscious Society" and "On Socialist Ecological Civil Society," among others. Written with the author's penetrating perspective, all these articles are profound explorations of environmental thoughts and theories.

The science-based and rational development approach aimed at building a harmonious socialist society are crucial to informing the new generation of Chinese leadership. Pan's articles, laced with strategically profound thoughts and innovative ideas about environmental protection, have developed and enriched thinking on green issues, and taken a central place in guiding environmental protection efforts.

Among Pan's articles published in 2006, "The Harmonious and Environmentally Conscious Society" and "On Socialist Ecological Civil Society" are the most influential.

II. *"The Harmonious and Environmentally Conscious Society"*

In this essay, Pan explains the conjoined concepts of the "harmonious society" and the "environmentally conscious society." From these definitions, and from descriptions of the relationship between these two things, he teases out various implications. He argues that the ideological principles and cultural values needed to broaden and deepen environmental consciousness should be strengthened and clarified. Supporting efforts to craft effective regulations and institutions makes this desire concrete. In particular, Pan proposes "rebuilding the strategic goal of creating an environmentally conscious society."

He believes that four interrelated actions need to be taken to fulfill this goal: incorporating environmental factors in macroeconomic decision making; reforming land use planning; establishing cradle-to-cradle[5] economic production systems; and formulating a new energy strategy.

Pan Yue defines a harmonious and environmentally-conscious society as follows:

> A harmonious society exists when all the elements of a society coexist, coordinate with, and promote each other. Taking this one step further, the *socialist* harmonious society refers to the state in which all persons in the state devote their talents to developing and maintaining this state, and contribute their effort toward those activities to which they are best suited. This brings about an even more complete sense of harmonious living.

Regarding environmental consciousness, Pan states:

> A society fully aware of the environment understands and works within the natural order. As such, individuals grasp the full meaning of natural carrying capacity and operate in a way that respects resource constraints. This society and the individuals within it will advocate for and work toward establishing harmony between man and nature, different individuals,

[5] Translator's note: "Cradle-to-cradle" economics refers to the reuse-and-recycling-based approach to design and production of goods. This approach stands apart from the more traditional product life cycle that follows a "cradle to grave" trajectory. The ultimate goal of cradle-to-cradle economics is to generate zero waste by having outputs of one productive system become inputs in another.

> groups, and organizations. Because of this, such a society adopts sustainable social, economic and cultural policies. China can adopt this consciousness to reduce its consumption in production, architecture, and housing first and foremost. This shift in viewpoint can move China toward sustainable resource usage and cradle-to-cradle design and production systems. This would propel China's economy forward through efficiency improvements, ongoing technological innovations, and transparent and liberal trade and financing systems. Taken together, these elements will enhance economic and social stability, while supporting the creation of a liberal and progressive socialist democracy.

Pan argues that environmental consciousness is subordinate to establishing a harmonious society. Environmental consciousness serves as an expression of the goal of building a harmonious society. It should be clear, however, that the two are interrelated.

Some people think that conservation-mindedness is similar or even equivalent to a fully developed environmental consciousness. Pan notes that conservation-mindedness stems from environmental consciousness.

In Part I of the article "What is a harmonious society?" Pan lays out four different perspectives on this point: the traditional Chinese view; the capitalist approach; the Marxist doctrine; and the Communist Party of China's point of view. He goes on to explain that building a Socialist harmonious society is the basic goal of the Communist Party of China (CPC). The CPC's methods rest on an eclectic combination of the best of the elements describe above. The essentials of Marxism coalesce with basic Chinese history and thought, as well as the broader sweep of human social evolution and scientific insights, to form this view and its expression in concrete terms. The approach is at once hybrid and singular in the way it melds traditional Chinese culture, Western civilization, and socialist insights and practice.

He goes on to encompass political theory in this essay. Of particular note is his observation that the CPC's promotion of Socialist societal harmony indicates that it has matured beyond revolutionary rhetoric and action. The CPC recognizes that social and political struggle does not represent the whole of Marxism; struggle is only one dimension of a larger evolutionary process. He advances the notion that building a Socialist harmonious society was Marx's lifelong pursuit, as well as the ultimate goal of Marxism. While Socialist movements continue to criticize Western industrial civilization as the means to advance its cause, the CPC takes a different tack. By building a Socialist harmonious society, the CPC uses tools from all Socialist movements as references, but transcends all of them by incorporating a wider set of tools and uniting them. In essence it has realized a more highly evolved form of Marxism.

In Part II, "What is an environmentally-conscious society?" he analyzes the concept's origins and implications. He then examines the relationship between an environmental consciousness and the realization of the harmonious society. He concludes the section by propounding the necessity for building such a society.

Part III of the article discusses how to construct an environmentally-aware society. He opens by proposing a complete update of China's cultural and value system. Based on this, he recommends that strategic development evolve in harmony with these new and more integrative values. In order to fulfill these goals, he suggests that various economic, social, political, and institutional mechanisms come into play to support and guide this effort. Finally, he makes recommendations to implement these changes comprehensively.

Within this analysis, he brings together several of these ideas in a potent statement:

> We should incorporate environmental factors in macroeconomic decision-making procedures from the local to the national level. A truly comprehensive analysis should include the five major resources: energy, minerals, fresh water, arable land, and flora and fauna. Many of these resources are in short supply and such shortages will grow more acute over time, finally straining carrying capacity in many cases. Large-scale developments and construction projects must consider these factors in the planning and implementation stages. Furthermore, this must fit into a broader and deeper effort to integrate resource and carrying-capacity analysis at the level of strategic economic planning, from considerations of productivity and efficiency to fundamentally transforming the economic production system. In particular, wasteful, high-consumption, badly organized economic activity aimed at achieving narrow short-term returns should be set aside in favor of thoughtful, well-managed, and sustainable developmental systems.

III. *"On Socialist Ecological Civil Society"*

In this article, Pan reviews the development of socialism in China and other countries. He illustrates and analyzes in detail Western eco-Socialist theory, which originated in the 1970s, pointing out its contributions to the countries that have put it into practice.

Pan defines the term *eco-civilization* as "the combination of material and spiritual achievements made by human beings under the principles of harmonious development and harmonious relations between individuals, society, and Mother Nature." He links this conceptualization to culturally-based ideas, such as "harmonious existence" and balance, productive and healthy exchange in social and economic dimensions, and comprehensive and systemic development, which lead to sustainable prosperity for individuals, organizations, societies, Mother Nature, and the bonds that connect them.

He grounds this analysis in a broad historical view. Through material pursuits, humankind has advanced civilization, at least in part. This process has been with humans from the dawn of civilization. However, limited natural resources, and/or limited technologies used to transform those natural resources into usable forms, have constrained material accumulation and wealth. Even though we have used technology to ease resource constraints, they have not altogether disappeared. Underneath economic growth, increased productivity, and technological improvements, many resources grow ever scarcer. Rather than pushing against resource constraints and violating systemic carrying capacity, individuals and society must learn to redirect their attention and energies from the accumulation of material fortune toward a meaningful, even spiritual, life. This can fulfill the aim of civilization and development. Taking this course will fundamentally transform human society.

Pan continues this part of the article with an examination of the internal consistency between traditional Chinese values and ecological demands. He underscores his thinking as follows:

> Chinese civilization entered industrialization late. Even though it is in the midst of this process, the society's basic spirit remains in accordance with ecological thinking and values, from the political system to its culture and even its underlying philosophy. Ecological thinking is near the core of traditional Chinese civilization. This makes it possible for us to reflect on and transcend the material civilization that has dominated mankind since the European Renaissance. By recognizing this linkage, China can take the lead in responding to ecological demands, eventually developing a truly ecologically sensitive and engaged civilization.

In the final part of this article, Pan looks ahead to envision what this society might look like. Rather than analyzing his point, it is more useful to let his words ring forth:

As far as China is concerned, the principles of ecological harmony offer a deep philosophical foundation for creating an ecological society. The Communist Party of China has put forward several decrees that underscore the importance that the government and the people attach to this process. The science-based and rationally organized outlook on development has been moved front-and-center in policymaking. Building a harmonious society, which includes raising environmental consciousness, underpins these efforts. All three efforts draw upon and draw together eco-socialism, sustainable development, and traditional Chinese values. This melding can lead us to fulfilling the aims of eco-socialism in China. This will revitalize the Chinese nation and restore socialism as the basis for organizing economic activity and social development. Because so much can be gained, China should meet the challenge courageously.

IV. *Repercussions of Pan Yue's Articles and Thoughts*

A. *Scholarly Commentary*

The publication of "The Harmonious and Environmentally Conscious Society" and "On Socialist Ecological Civil Society" stirred strong reactions in China and abroad. Although they have provoked criticism, they have been well received in general.

Many well-known scholars have responded to Pan's articles positively. Professor Wei Jie, from the College of Economic Management, Tsinghua University, said, "'The Harmonious and Environmentally Conscious Society' makes a comprehensive analysis of how to construct an environmentally aware society." He praises the article's breadth and completeness in the way that it unites theory and policy systematically. Ideology, societal considerations, and strategic aims all blend seamlessly in the analysis.

Zhou Fengqi, a researcher at the National Development and Reform Committee (NDRC), points out the strategic value of the work, particularly how it penetratingly unlocks the linkage between society and nature through the lens of science-based rational analysis.

Han Deqiang, a researcher from the Management College at Beijing University of Aeronautics & Astronautics, approaches the article based on three perspectives. First, it combines the definitions of the harmonious and environmentally-conscious society, providing a solid point of origin for building the harmonious society. Secondly, it combines traditional Chinese values with Marxism. Third, it fuses the development of the harmonious society with the entire economic and political reform process, particularly the new emphasis on democratic development. This analysis points the way toward a new body of reforms.[6]

Professor Wang Canfa from the Institute of Environmental Recourses Law at China University of Political Science and Law points out that the article

> illustrates the theory, policy and mechanisms for building an environment-conscious society in four aspects: the ideological, the civil, the strategic, and finally, the systemic. He clearly envisions what "harmonious society" and "environmental consciousness" means. This shines new light on these points for readers. Through this body of work, the understanding of traditional Chinese values, Marxist values, and their relationship to one another, is deepened.[7]

Gu Haibin, a professor at Renmin University, also thinks highly of Pan's two articles, saying one can identify with Pan while reading the article. He says that these articles are convincing because of their strong arguments based on reliable and well-organized facts and data.

Jia Kang, Dean of The Research Institute for Fiscal Science affiliated with the Ministry of Finance, opines that Pan's two articles are clear, open-minded, and predicated on brilliant logic and insightful analysis.

Interestingly, these by-lined commentaries mainly appeared on some major websites rather than in traditional media. In general, commentary on Pan's articles has appeared online.

Many critics agree that Pan's theory on eco-socialism draws upon socialist ideology while making original modifications. They argue that the theory on eco-Socialist civil society is based on the principal of practicality. It struggles with and answers contemporary questions, which enriches Socialist ideology to a great extent. Commentators add that such theoretical achievements address the real-world situation in China. Even though Pan's analysis is grounded in history, it still raises new points of view, bringing the theory of environmental protection to a higher level.

---

[6] Han Deqiang, "Can an Environmentally Conscious Society Be Realized?" (July 5, 2006), http://www.people.com.cn.

[7] Wang Canfa, "My Opinions on 'The Harmonious and Environmentally Conscious Society,'" (August 23, 2006), http://www.cust.com.cn.

Both of Pan's articles concern new understandings of socialism. Critics think that there are several points of view in his articles that should be noted: first, that the pursuit of harmony is the essence of socialism. Repeatedly, Pan argues that the final expression of socialist ideology is harmony rather than struggle. Second, the most important theory expounded in "On Socialist Ecological Civil Society" is the explanation of the relationship between the concept of science-based, rational development and socialist ideology, which is rarely seen in other articles on socialism and/or development. From his point of view, only when a sustainable market-based production system evolves into a science-based development system can such a productive system offer lasting and comprehensive gains to Chinese society. Over time, this evolution will even extend beyond China. Third, we must confront European-style eco-socialism. In Pan's opinion, eco-Socialism transcends and supercedes watered-down and piecemeal European efforts. True eco-Socialism, as the name implies, unites ecological thinking and action with Socialist thinking and action. This new innovation lies at the heart and is the ultimate realization of the Socialist agenda. He illustrates this linkage systematically, pointing out that ecological civil society can only be achieved through Socialism; capitalism is simply incapable of producing the values, balance, and completeness needed to realize eco-Socialism. However, Socialism can only fulfill its ultimate mission by integrating a full range of ecological values.

B. *The Value of Theory in Policymaking*

New theories typically provoke new thinking, and the field of development is no exception. In the process of deepening and expanding reform in China, conflicts and problems have emerged. Dealing with such deep-seated issues requires that thorough research be carried out before attempting to implement any solutions.

For instance, tackling environmental problems requires us to think systemically and develop comprehensive solutions. A variety of crucial problems with local, national, and international implications need to be dealt with in this fashion. Applying Socialism in this process of systemic analysis and problem solving arises directly form Pan's finding that "Only through socialism can we achieve ecological civil society."

Besides being a government official, Pan Yue is a poet, a fighter, and a thinker. In 2006, his articles reached a new level of profundity.

His deep thinking has opened up a new space for further research on environmental theory in China and beyond. He has pointed a way forward for environmental thinkers, Socialists, and the whole of society. Perhaps the most valuable thing we can learn from his creative thinking is the spirit of creativity itself.

CHAPTER FOUR

ENVIRONMENTAL FISCAL REFORM (EFR) IS THE KEY TO REALIZING ENVIRONMENTAL TARGETS IN THE ELEVENTH FIVE-YEAR PLAN

Ma Zhong and Wu Jian

The Eleventh Five-Year Plan for the National Economy and Social Development clearly defines environmental targets, such as "reducing pollution." It also states that "restrictive forces" can be used in meeting these targets. This article examines the feasibility and problems in achieving these aims by analyzing capital demands and other economic pressures that may undermine such efforts. Problems with the current financial mechanisms used are uncovered. The article then argues that employing Environmental Fiscal Reform (EFR) would be a superior policy for realizing environmental policy goals stated in the Eleventh Five-Year Plan.

Key words: Environmental Fiscal Reform (EFR), financial mechanisms, environmental protection

I. *Capital Pressures in Meeting Environmental Targets in the Eleventh Five-Year Plan*

The Eleventh Five-Year Plan for the National Economy and Social Development has set the central environmental target for the following five years: achieving a 10 percent reduction in total discharge of major pollutants by 2010. This is the first time that the Chinese government has stated an environmental protection "obligatory target" in a Five-Year Plan. This statement not only carries legal force, but also clearly states that these targets will "strengthen government's obligation [to environmental protection efforts]. This policy will be carried out by rationally allocating public resources and efficiently using the administrative power of the government." Obviously, the environmental targets will play an important role in the government's work in the following five years.

China's environmental targets are becoming clearer and more rigorous. These changes provide evidence that environmental awareness is growing and making itself felt through stronger political demands. However, since China failed to reach the Tenth Five-Year Plan's environmental targets, it is too early to be sure that the even stronger environmental targets in the Eleventh Five-Year Plan will be achieved. For instance, by the year 2010, sulfur dioxide and chemical oxygen demand (COD) emissions should be down 10 percent from 2005 levels. On average, emissions should fall 2 percent per year, or 1 percent every six months, to meet this goal. Unfortunately, current trends show emissions climbing rather than declining through the first half of 2006: COD and sulfur dioxide emissions reached 6.896 and 12.724 million tons.[1] These figures mark increases of 3.7 percent and 4.2 percent in just the last year.

The reason for this increase is simple: China's prodigious economic growth. During the first three quarters of 2006, China's GDP climbed by 10.7 percent compared to the same period in 2005. The GDP jumped by 11.3 percent in the second quarter alone. Growth will exceed 10 percent for the totality of 2006.[2]

Rapid economic growth continues to be driven by investment in heavy industry, which consumes vast sums of energy while emitting huge quantities of pollution. For example, electricity generated by thermal power plants rose by 12.5 percent between January–July 2005 and January–July 2006. This, in turn, increased sulfur dioxide discharges. COD output rose because of accelerating urbanization, increased wastewater discharge, and incomplete construction and installation of planned wastewater treatment plants and piping systems. One expert from the State Environmental Protection Administration (SEPA) of China estimated that China's continued double-digit GDP growth rates will utterly undermine energy conservation and environmental protection goals unless special measures are taken to improve operational efficiency and emission treatment systems.[3]

It is disappointing to see that environmental protection policies remain the plaything of economic growth and industrial policy. Improving environmental conditions requires greater investment and that investment may well be paid for through growth. In other words, the need to balance growth, environmental conditions, and capacity constraints

[1] Translator's note: Note that by "tons" we mean metric tons, i.e., one thousand kilograms, or about 2204.6 pounds in English measure.

[2] The Chinese Statistics Bureau, *An Annual Increase of Over 10% Is Ensured in 2006* (2006), http://focus.jrj.com.cn/news/2006-10-30/000001742327.html.

[3] Ding Jun and Wang Shiling, "Campaign to Defend Environmental Targets for the Eleventh Five-Year Plan," *21st Century Economic Reports*, September 17, 2006.

poses a dilemma. Clearly, China has yet to adequately address this problem.

II. *Rethinking China's Environmental Protection Financial Mechanisms*

Failure to reach environmental targets during the Tenth Five-Year Plan was caused by several factors, particularly underinvestment in technical improvements. The State Environmental Protection Administration publishes data on environmental conditions and progress made toward meeting various environmental protection targets. These figures show that during the Tenth Five-Year Plan, major environmental projects suffered a variety of problems. Technical improvements, such as the installation of desulphurization equipment in power plants, and various pollution control and prevention measures aimed at improving conditions in key rivers, fell well short of targets. By the end of 2005, only 70 percent of the planned reduction of sulfur dioxide discharges had been achieved. Of 256 projects planned to prevent acid rain and sulfur dioxide pollution, only 54 percent has been completed and put into operation. Only 1,378, or 63 percent, of the 2,130 pollution control projects listed in the Tenth Five-Year Plan reached completion by 2005. The total investment in environment protection projects equaled 86.4 billion RMB, or about 53 percent of the planned level of investment. By the end of 2005, numerous water treatment and restoration projects remained incomplete: 30 percent of the projects on the Huaihe River; 44 percent of those on the Haihe River; 57 percent on the Liaohe River; 14 percent at Taihu Lake; 47 percent at Caohu Lake; and 46 percent at Dianchi Lake (three rivers and three lakes).[4] These statistics underscore the problem of underinvestment, but also the problem of developing systems for meeting stated goals.

This begs two fundamental questions: Who is responsible for investments in environmental protection, and why have they failed to fulfill their obligations?

Environmental protection investments can be divided into private and public categories. Although the legal principle that "those who

[4] The Chinese Institute for Environmental Planning, "Analysis of the Completion of Environmental Targets Set by the Tenth Five-Year Plan," SEPA press conference, April 12, 2006.

pollute the environment are responsible for its treatment" clearly defines responsibilities for environmental investment, we still find that the investment from government is of great importance, especially regarding pollution prevention. The scale and efficiency of private investment depend on government investments and policies to a large degree; in short, the government's investments and policies play a special role in the environment protection market.

Government plays a singular role by using coercion to move policies and policy implementation forward. Large and often concentrated investments enable governments to monitor and regulate environmental protection efforts, including those made through non-government and private investments.

The government also guides and shapes investment by creating favorable conditions to support policy aims. For instance, subsidies and taxes can steer investments in particular directions.

Finally, the government balances investment efforts across various dimensions: geographical, social, economic, and ecological. Of particular importance is balancing investment across different regional, provincial, and local needs.

Therefore, government investment—that is, expenditures—directed toward environmental protection should be given primary attention. Through these investments, government performs its duty of protecting the environment, while encouraging and focusing the scale and direction of total investment devoted to environmental protection.

A. *Inadequate Government Investment and Insufficient Environmental Supervision*

Government cannot adequately implement its regulatory function without first developing the necessary capacity to carry out monitoring, supervision, and enforcement functions. Developing this capacity should be a central goal since environmental protection is one of the basic state policies in China. The legal structure for these policies rests on the 1979 *Environmental Protection Law of the People's Republic of China*. Even though this law grants the government special powers and responsibilities regarding environmental protection, the government has yet to specify its environmental protection expenditures in the budget. That is, environmental protection efforts do not enjoy stable, reliable financial support. This is quite unusual; leaving aside such budgetary considerations undermines the coordinated development of environmental investment needed to deal with the side effects of economic growth.

Unstable and inadequate funding has affected the development of many environmental protection departments, especially in the localities. Environmental supervision will remain inadequate in the face of such funding shortfalls. In this regard, 2005's Songhua River Toxic Spill Crisis sounded an alarm. Lack of capacity can be seen in every step of this disaster: government failed to adequately supervise firms generating pollution; the environmental protection department showed weakness in dealing with this and similar emergencies; and information from water quality monitoring efforts is both inaccurate and generally unavailable for inspection. Given lax conditions and the lack of capacity, it seems likely that similar pollution accidents will occur in the future.

Regionalism further exacerbated these problems. Forty-five thousand people now work in environmental protection; thirty-eight thousand of these work at the county or district level. Under the current environmental protection managerial system, local environmental protection departments are ruled by local governments and financially dependent on them; the enforcement of environmental protection laws largely depends on the financial status of these localities, as well as their level of commitment to environmental protection. Since most local governments place economic growth at the top of their policy priorities, and because many policymakers see growth and environmental protection as being in conflict, environmental protection measures and departments remain systematically underfunded at the local level.

B. *Inefficient Use of Investment for Environmental Policies*

1. *The Need for Government to Lead Based on Science, not Markets* Investment levels have not been adequate to meet environmental needs in spite of the pivotal role played by government. A key sticking point continues to be the government's strong tendency to follow the market.

Not only have funding levels been inadequate for the task at hand, funds have not been directed in a way that generates the greatest impact. For example, the central government invested 53.4 billion RMB from 1998–2005 in the construction of urban infrastructure designed to improve wastewater treatment. However, data show that systematic and comprehensive wastewater treatment has yet to be realized. Funding has been spent irrationally, quite possibly in response to political pressure rather than engineering and ecological needs.

Investments are mainly concentrated in cities located in the lower reaches of major rivers. These tend to be economically-developed large cities in eastern coastal areas. On the other hand, cities upriver, especially in the western and less developed provinces, have very low wastewater treatment rates. For instance, in Shandong Province, which includes the Yellow River delta, 43.46 percent of its wastewater receives treatment. Upriver in Shaanxi Province, the rate falls to only 20.63 percent. Note that since both provinces lie on the same river system, the treatment rate upriver fundamentally affects water conditions downriver. For the river to receive adequate treatment, efforts must be applied systemically. This means that more resources must go into treatment of upriver waters. However, political and economic pressure pushes resources downriver, where the overall impact of wastewater treatment is fundamentally diminished.

Things may actually be worse than what these low officially-reported treatment rates reveal. The national television network China Central TV (CCTV) produced an investigative report showing that the actual treatment rate is less than 15 percent.

Market forces will tend to reinforce these shortcomings. Therefore, the government must push investment based on ecological and engineering needs rather than in response to market and political pressures.

2. *Investments have not Achieved Broad Effectiveness* In the past several decades, some environmental infrastructure has begun to take shape. However, these resources have not been efficiently used, which has significantly reduced their impact. Underperformance has produced a secondary impact: it reduces the incentive to make further investments in environmental protection infrastructure.

This unfortunate state arises from various factors. For instance, the development and implementation of these systems generally lack effective supervision systems. Furthermore, economic incentives and pricing policies do not support environmental investment objectives. For example, electricity prices are not rational, given demand levels. This has lead to overuse of coal and overbuilding of coal-fired power plants. Artificially low costs of operating a coal power plant, and the low cost of sanctions for environmental violations, practically guarantee that sulfur reduction equipment will not be installed, used, or maintained.

In 2004, desulphurization had been implemented in a small number of coal-fired power plants in China; only about thirty million kW being produced could be considered "clean" coal energy.[5]

C. *Investment in Environmental Protection Falls Short of Demands*

To be effective, environmental protection efforts require steady and long-term investment based on the rate, scale, and pattern of economic growth. However, these basic factors have not been taken into full consideration when determining the scale and kinds of investments.

1. *No Reliable Source for Capital Investment in Environmental Protection Exists*

Funding for environmental protection efforts, including capital development and infrastructure projects, remains unreliable since allocations are vulnerable to political maneuverings and legislative fiat. Even so, spending levels have increased. Unfortunately, virtually no strategic plan or inter-organizational coordination exists. This has led to the irrational allocation of funds across a motley assortment of departments and projects. Increased spending may have exacerbated the problem.

Funding shortfalls have been particularly acute in some departments. Remarkably, departments with national security and long-term interests and responsibilities have been among these. In particular, funding needed for the enforcement of environmental laws, trans-provincial water treatment, and national nature-reserve administrative oversight top the list. A clutch of other needs face similar difficulties, including implementing international environmental agreements, constructing nuclear waste disposal facilities, and treating large and unresolved pollution

[5] Translator's note: A back-of-the-envelope calculation can illuminate the scope of investment and its impact. If the costs of constructing and operating the desulphurization equipment are 1.5 cent/kW and an average of fifty-five hundred hours of electricity is generated every year, then the annual cost for desulphurization would be 2.475 billion RMB. If all these facilities, with a total worth of 8–13.4 billion RMB (calculated by the construction fee of 300–500 RMB yuan/kW) are taken at full use, the SO2 discharge volume will be reduced by 1.8 million tons every year, or about 70 percent of the target set by the Eleventh Five-Year Plan. Moreover, the pollution discharge fee will be cut down by 1.08 billion RMB every year (0.6 yuan/kg SO2), and 36 billion RMB of loss caused by environmental damage will be saved. This calculation shows that the price of electricity coming from plants that have implemented desulphurization can play an important role in bringing the invested environmental resources into full play, which can create huge economic and environmental benefits.

cases. Ignoring this problem will threaten China's continued social and economic development along with its environmental security.

2. *Funding Allocation Focuses on Narrow, Short-Term Goals, Especially Emergencies*
By focusing funding on departments and projects with a narrow remit and crisis management, spending fails to take on a coherent, comprehensive, systemic character. Efforts lack consistency; sometimes outright wastefulness results. For example, spending on efforts to "return farmland to forest" was motivated in part by rural fiscal reform rather than environmental protection policy.

3. *Environmental Protection Spending should be Based on Economic Performance*
Budgets for environmental protection efforts should be based on the speed and scale of economic expansion. Spending on environmental protection has failed to keep pace with rapid economy growth in the past twenty years. This has left such efforts under-resourced relative to the growing scale and complexity of environmental impacts produced by economic activity.

During the first three Five-Year Plans, covering the years 1953–1970,[6] 2.17 percent, 2.46 percent, and 2.37 percent of GDP was invested in environmental protection. By the years of the Eighth, Ninth, and Tenth Five-Year Plans—covering 1991–2005—investment rates had fallen to 0.74 percent, 0.88 percent, and 0.99 percent, respectively. Premier Wen Jiabao admitted that inadequate investment in environmental protection was one of the major reasons for ongoing shortcomings in meeting environmental targets set by the Tenth Five-Year Plan. Investment in environmental protection in the Eleventh Five-Year Plan is slated to increase 85 percent over expenditures made in the Tenth Five-Year Plan. Note, however, that this increase does not keep pace with GDP growth, which gallops forward by 80–120 percent every five years.

D. *Improper Pricing and Fee Assessments Thwarts Investment*

Various characteristics of the environmental protection market dampen private investor interest. Ineffectiveness and inefficiency in the use of

[6] Translator's note: The first three Five-Year Plans are: First Five-Year Plan (1953–1957), the Second Five-Year Plan (1958–1962), and the Third Five-Year Plan (1966–1970). Note that during 1963–1965, China entered a protracted planning process that delayed the launch of the Third Five-Year Plan.

funds currently allocated turns away many investors. Moreover, environmental NGOs (ENGOs) have never played a sufficiently strong, clearly-defined role.

Returning to the example of urban wastewater treatment highlights these problems. As noted earlier, for water quality across a river basin to improve, efforts must be focused on upriver pollution. In spite of this obvious fact, funding in upriver areas falls well short of the sums flowing to downriver locations. A few figures regarding fee rates underscore the extent of the problem. Well-developed cities in the east like Nanjing, Shanghai, and Beijing set their treatment fees at 1.00 yuan, 0.9 yuan and 0.9 yuan, respectively. Less developed inland cities like Xi'an, Lanzhou, and Xining set the corresponding fees at 0.36 yuan, 0.3 yuan, and 0.27 yuan, repectively Awash in funds, eastern cities can pay for more wastewater treatment, and thus attract far higher levels of investment in treating the problem.

Funding allocation is only part of the problem; price distortions also create systemic dysfunction. The current policy sets pollution discharge fees so low that firms have virtually no incentive to invest in pollution controls. Thus far, the government has been unwilling or unable to adjust taxes and prices to provide the impetus to spur environmental protection investment.

III. *The Significance of Environmental Fiscal Reform (EFR)*

Environmental fiscal reform adjusts the relationship between economic development and the environment. These fiscal measures touch upon many areas, from the protection of natural resources and ecological systems to the provision of services that ensure public safety through environmental protection. Note that fiscal decisions depend on more than collections and expenditures; pricing plays a key role in financing environmental initiatives.

A. *Adequate Finance is a Necessary Ingredient*

Fiscal measures play an important role in environmental protection efforts. Meeting stated targets depends crucially on adequate funding. Without it, goals and targets are just so much hot air.

B. *Regulating the Market*

Finances structure markets at the most basic level. The level of funds devoted to any effort indicates the government's commitment to that policy. Moreover, government funding supplements private investment. Typically, government carries out functions that the market will not. In public goods investment, which covers many environmental resources, government may well be the only investor.

Even as the government bears burdens that private operators do not, it has tools at its disposal that sets it apart. Most obviously, the government can raise revenues through taxes and price adjustments. Prudent and thoughtful use of taxation and other tools can improve resource usage and allocation. This addresses environmental problems in the most fundamental and comprehensive way.

C. *Effective Management of Government Operations and Revenues*

The needs demanding government attention inevitably overwhelm government revenues. To ease the impact of this condition, government must operate efficiently. As part of this effort, administrative and financial powers should be linked and work in harmony. Unfortunately, because it is a relatively new government function, environmental protection lacks a strong political base, which undermines its administrative capacity. Better-established and more widely-appreciated public sector activities enjoy greater support, including education, public health, and agriculture.

In order to meet its obligations, government must redouble its efforts in raising revenues for environmental protection, use those funds effectively, and create a support base for environmental policies so that support will continue and not be subject to wide fluctuations in financial resources and spending priorities.

D. *Creating a Market*

Policy has created and shaped the environmental protection market. The measures arising within this market should encourage the redeployment of resources according to environmental needs, regional differences, urban and rural distinctions, and systemically related ecological demands.

As with any market, revenues and costs determine the market's scope. In the case of wastewater treatment, establishing the market depends crucially on collecting usage fees from those disposing of wastewater. By developing policies and fee payments that consider regional needs, systemic relationships (especially for river systems), and scientific analysis, the currently-applied policies can be replaced in favor of a more effective regime. Specifically, the concentration of funds and resources in the easterly, downriver cities will give way to a new system that will concentrate funds and treatment in upriver areas. This will restore the health of many more miles of river, while improving access to clean water for millions. It will also reduce pressure on downriver treatment facilities.

Note that accomplishing this depends crucially on fees. In turn, fees should be based on the true cost[7] of resource use, as well as the total costs for maintaining the health of environmental systems at reasonable standards. Nowhere is this more important than in water quality standards. While a market-based system will not yield utterly pristine water, it will, with proper structure and prudent intervention, improve water quality standards necessary to maintain agriculture, aquaculture, and drinking water. Note, as well, that a well-thought-out system will not require any one region to pay exorbitant fees. This problem has cropped up in some areas, and it seems driven by the desire of some individuals and organizations, including companies, to enjoy extraordinarily high profits on environmental projects. Government will have to play an especially large role in seeing that otherwise neglected areas, or areas that do not have ready access to needed revenues, still receive the resources needed to build and maintain effective wastewater treatment facilities.

---

[7] Translator's note: "True cost" refers to a system of cost accounting that seeks to integrate externalities, such as pollution and social costs, into standard economic calculus. A number of economists have been calling for this shift, including Kenneth Boulding, Herman Daly, and Robert Costanza. True cost economics has also been strongly associated with so-called "triple bottom line" accounting, which considers people and the planet, alongside the more traditional economic measure of success and profit. Consultant, author, and social entrepreneur John Elkington coined the phrase in 1994.

IV. *Environmental Fiscal Reform and Environmental Protection Goals in the Eleventh Five-Year Plan*

With the economy booming and government revenues up, the time is ripe for solving environmental problems.

Since 1998, the Chinese government has implemented financial policies that have encouraged rapid, but stable, economic expansion. GDP reached 11,669.4 billion RMB in 2003, or four times the 1993 level. Governmental revenues climbed by double-digit rates in these years, and actually outpaced overall growth. In 2003, total government revenues equaled 2171.5 billion RMB, up nearly 500 percent since 1993. The central government took in 1246.5 billion RMB of the total. This sum exceeded the 1994 total, when the government reformed the tax system, by more than 400 percent.

Momentum continued beyond the twenty-year mark. In 2005, total government receipts topped 3162.798 billion RMB, up 523.151 billion RMB over the 2004 level. In other words, revenues grew a whopping 19.8 percent in a single year. Total receipts during the Tenth Five-Year Plan equaled 11,500 billion RMB. Both central and local governments enjoyed growth in revenues that more than doubled total receipts collected during the Ninth Five-Year Plan. National revenue accounted for 17.3 percent of the GDP.[8] Armed with greater resources, the government can now move aggressively to implement reforms and tackle lingering problems.

A. *Clarifying Responsibilities in Meeting Environmental Protection Targets*

Money can help provide the resources needed to fulfill environmental protection targets. However, the proper use of funds depends on properly distinguishing the roles that markets will play from those played by government. In turn, the functions taken up by the central government must be distinguished from those better addressed by local governments. Making these distinctions and basing sound decisions on them will help direct funds where they will have the greatest impact.

[8] Ministry of Finance of the People's Republic of China, *Report on the Draft Resolution of the Implementation of the Central and Local Budget for 2005* and *the Central and Local Budget for 2006*, The Fourth Session of The Tenth National People's Congress.

Environmental protection programs in the Ninth and Tenth Five-Year Plans had to make due with funding based on a percentage of GDP. However, this approach said nothing about the different roles played by local and central governments, or about the role of the public sector as distinct from the private sector. Problems with collecting data from companies further aggravated these problems, since the full scope of the need could not be measured, nor could an accurate assessment of private-sector spending be calculated.

Even though private companies have relatively clear guidelines to follow, they take their cue from the government's environmental efforts. Furthermore, the government's role as supervisor of private efforts must be adequately supported in order to function properly over time.

Ultimately, jurisdiction and investment in these efforts should be based on a scientific assessment of environmental problems and their scale. Local governments should solve particular and specific problems; the central government should be responsible for environmental problems of transregional or national scope.

B. *Stabilizing Environmental Expenditures*

Fiscal policy and the complexity and scale of the problems facing the government demand that environmental expenditures get their own line item in the budget. This will bring coherence to spending. It should also stabilize funding levels and make the distribution of funds more predictable. This step will be written into the Budget and Final Accounts of the Central Government (2006), and implemented in 2007. To achieve broader coherence in budgeting and spending, however, spending through various departments should be consolidated and/or coordinated so that the total scale of spending can be more easily determined. This will help policymakers achieve a better match between the scale of spending and the scale of the demands that need to be met.

Because dealing with environmental problems requires a long-term commitment with uncertain payoffs, at least in the short run, funding must reflect a long-term and stable commitment to effectively treating problems. This demands that funding be stable and predictable, which should prod fiscal managers to decouple environmental protection funding from overall fiscal and macroeconomic fluctuations.

C. *Setting Prices to Achieve Desired Outcomes*

Price fixing is one of the most powerful and effective tools that the government wields in its efforts to manage markets. Using it can help to compensate for various market failures or price distortions generated by markets.

Price setting can be used to produce change in those markets which most need it. For instance, in generating "cleaner" energy—that is, energy that comes from power plants using sulfur reduction technology—adjusting prices can encourage operators to install emission-reducing technologies. As it stands, prices for coal and electricity encourage the use of high-sulfur soft coal and barebones technologies in the design and operation of power plants. This situation can be changed through price adjustments. By carefully monitoring inputs and outputs, and by setting emission and energy output targets based on carrying capacity, the price system can be used to guide operators toward a sustainable level of sulfur output, among other emitted gases.

In order to move protection efforts forward, other tools can be used. Tax incentives and subsidies can produce change. Ideally, these should be used during transitional periods; once market mechanisms with robust and rational pricing come into play, other mechanisms should be removed.

D. *Tax Policy Reform within the EFR Framework*

Reform of the tax system should focus on changing economic activities rather than generating new revenues. Setting fee structures should be predicated on achieving desired environmental outcomes rather than fulfilling funding targets. As this process matures, various pollution discharge fees should be adjusted in line with the overall environmental financial expenditure and collections system.

Based on the best practices of other countries, a mature and functional public financing system would include a significant role for environmental taxation. China's fiscal system is not yet this developed. However, by establishing the elements needed in the reform process, and the building blocks of an environmental taxation system, China can begin moving toward a more robust and effective taxation system that addresses environmental needs. Researchers should take the long view and think strategically about how to achieve this change.

Such research should consider current scientific knowledge embedded in a thoroughgoing cost-benefit analysis. The effect of this move on environment and natural resource policies would be profound.

A variety of taxes need to be considered within such an analysis. Obviously, taxes on basic resources need to be front—and center in this research. However, less obvious taxes should be included, such as vehicle sales taxes and fuel taxes, which have a major impact on environmental outcomes.

As it sets increasingly rigorous environmental protection targets, the government should assume greater responsibility in seeing these targets met. This will require the government to more effectively monitor inputs. It will demand that market failures be addressed systemically. Externalities will need to be comprehensively tackled, ultimately leading to a policy that ensures those who produce pollution will take on the task and cost of treating its effects.

To reach these goals, environmental finance reforms need to be deployed sooner rather than later. By putting EFR into play, environmental protection schemes and economic development can be linked. A necessary step on this path is to provide the stable and adequate funding for meeting long-term environmental goals. Taking these steps is the key to realizing the environmental protection targets set by the Eleventh Five-Year Plan.

CHAPTER FIVE

A GOOD BEGINNING: ENVIRONMENTAL LEGISLATION IN THE ELEVENTH FIVE-YEAR PLAN—A 2006 UPDATE

Ning Chen and Wu Zhijiao

The year 2006 marks the first year of the Eleventh Five-Year Plan. This pivotal year connects advances made during the Tenth Five-Year Plan period to conservation and sustainability targets stated in the Eleventh Five-Year Plan.

Environmental legislation in 2006 addressed four themes: bringing democratic processes into environmental protection; developing market-oriented tools for environmental protection; strengthening legal monitoring and supervision; and crafting clear and detailed environmental regulations. These newly-enacted laws are meant to tackle new problems and respond to new developments. In short, the environmental legislative process was a huge success in 2006.

Key words: environmental legislation

I. *New Laws Tackle New Problems*

From January 1 to October 1, 2006, a number of local and national environmental laws and regulations were drafted and put into effect. These laws helped to fill in legal gaps while giving government a broader toolkit for tackling emerging problems. At the same time, local regulations moved beyond the vagaries of national environmental laws by including innovations born out of local practice and experience.

A. *Developing and Implementing the* Renewable Energy Law: *A True Great Leap Forward*

In the foreseeable future, China will face three dilemmas in energy consumption: the choice between using large amounts of coal or focusing on environmental protection, especially in relation to global warming; the pressure to use higher-quality energy sources in the face of increasingly acute shortages and falling oil and gas reserves; and the tradeoff between meeting energy demands with imports and maintaining energy security.

Freeing the country from these dilemmas will require a significant change in energy policy. Developing the renewable energy sector offers one of the best approaches for solving these problems. Unfortunately, China has dragged its feet in outlining concrete goals or timetables for developing renewable energy. No legal structure exists regarding renewable energy. Standards have not been developed. As a result, the sector has languished.

In 2003, interest in doing the groundwork needed to develop the sector began building momentum. By 2004, the *Renewable Energy Law* had been drafted. It passed in February 2005 and went into effect on January 1, 2006. A number of the law's features are worth highlighting here. These include the establishment of a total output figure for renewable energy, mandatory connection of renewable energy producers to the national power grid, development of pricing and fee structures for renewable energy, and the creation of a stand-alone fund for renewable energy development.

While these steps are significant, a very long march lies ahead. The effectiveness of the law depends on the implementation of complementary laws and regulations. The current task, therefore, is to put in place laws that detail the implementation process.

The *Renewable Energy Law* stipulates that supportive regulations and initiatives will be developed and implemented in the following twelve areas:

1. hydroelectric power development;
2. surveys designed to determine technical needs;
3. renewable energy output targets;
4. development and utilization plans for renewable power;
5. operational plans for renewable energy firms;
6. pricing policy for renewable energy, including the cost of power generation and grid connection;
7. fee-sharing arrangements for renewable energy power generation projects;
8. establishment of funds for renewable energy development;
9. set-asides for projects in rural areas;
10. preferential loan policies, including low-interest loans;
11. earmarks to encourage solar energy construction and use; and
12. technical standards development, including grid connection for renewable energy firms.

On April 30, 2005, the Office of the National People's Congress (NPC) Standing Committee issued an edict requiring departments with expertise in energy, pricing, finance, construction, and standardization to draft the aforementioned administrative regulations and technical standards. The National Development and Reform Commission (NDRC) followed this with a meeting on June 22, 2005, to lay the groundwork for this process. The NRDC will be the short-term point of focus in implementing the *Renewable Energy Law*.

Even though the law moves this process forward, it has some glaring faults. Chiefly, the goal for total renewable energy output is not sufficiently ambitious. Numerous scholars have pointed out that the *Renewable Energy Law* should encourage renewable energy in the short term to reduce fossil fuel use, develop renewable capacity in the midterm, and advance renewables to be the dominant—if not only—energy source in the long term. Mr. Huang Ming, the law's principal sponsor, described output targets as follows:

> According to the national renewable energy development plan, renewable energy will only account for 15% of the total energy consumption in 2020. The target is significantly lower than that of some foreign countries, which plan to draw 30%–40% from alternative sources by 2020.

Given these low output standards, renewables cannot serve as a substantial substitute for coal, oil, nuclear, and gas (CONG) energy sources.

The second major problem with the law is the absence of an accountability clause. That is, output targets are not binding. Chapter 7 of the *Renewable Energy Law*, which focuses on "legal responsibilities," does not include any legal basis for prosecuting those who fail to fulfill Articles 7 and 8, which pertain to total renewable energy output. Without sanctions it seems unlikely that the targets will be met.

Finally, the incentives described in the law clearly favor some technologies over others. Information on wave and wind energy is notably scarce. This may indicate that China lags behind the rest of world in research and development in these technologies, and further, that it has made no provisions for getting access to these technologies through trade, strategic alliances, joint ventures, or other technology-sharing agreements.

B. *Formalizing the Judicial Role*

In order to prosecute and punish environmental crimes, the Supreme People's Court issued *The Interpretation of the Law in Criminal Cases Involving Environmental Pollution* (hereinafter referred to as the *Interpretation*) on July 21, 2006, which was put into effect on July 28, 2006. The *Interpretation* offers detailed definitions of "severe public and private property losses,"[1] "the severe consequences in cases that involve personal injury or death,"[2] and "severe damage to human health," and promises "especially severe consequences"[3] for those found responsible for causing environmental pollution accidents as described in Articles 338, 339, and 408 of the *Criminal Law of the People's Republic of China (PRC)*. It also specifies the criteria for convicting and sentencing the guilty parties. As a result, the *Interpretation* will strengthen the effectiveness of the *Renewable Energy Law* (and related laws) in prosecuting environmental crimes. The *Inter-*

---

[1] According to the *Interpretation*, any of the following would be defined as "severe loss of public and private properties": losses exceeding 300,000 RMB in public and/or private properties; damage that hampers the long-term biological function of more than five *mu* (0.3335 hectares or 0.8235 acres) of farmland, forest, or other woodlands, or over ten *mu* of other farmland, or over twenty *mu* of other land; or that destroys more than 50 m^3 of woodlands or kills more than 2,500 saplings in a given area.

[2] According to the *Interpretation*, any of the following three classifications will be considered "severe damage to human health": events that cause more than one death, more than three severe injuries, or more than ten moderate injuries, or that cause more than one severe and more than five additional moderate injuries; events that cause infectious disease to break out and spread, or cases of poisoning determined to be a Grade III public health emergency, as described in national standards designed to prevent and/or limit public health emergencies; and finally, cases that produce "personal injury or death... [or] severe damage to human health."

[3] According to the *Interpretation*, any of the following seven cases shall be described as having "especially severe consequences": events that cause over one million RMB worth of public and/or private property damage; events that cause water pollution and demand population evacuation that would be classified as worse than a Grade II public health emergency as described by the national standards designed to prevent and/or limit public health emergencies; and events that fundamentally disrupt the fertility of farmland or result in long-term damage to farmland in excess of thirty *mu* (2.001 hectares or 4.941 acres) or of over sixty *mu* (4.002 hectares or 9.882) of other land; or events that cause the death of over 150 m^3 of forest or other woodlands, or the death of over seven thousand five hundred saplings; or events that cause more than three deaths, more than ten severe injuries, more than thirty moderate injuries, or that cause more than three severe injuries and more than ten other moderate injuries; events that cause infectious disease to spread until reaching a Grade II of public health emergency as described in the national standards designed to prevent and/or limit public health emergencies; or events that produce especially severe consequences not described above.

pretation is a milestone in China's environmental legislation since it is the first environmental law with unambiguously articulated clauses regarding standards of evidence and guilt, with straightforward guidelines for putting these assessments into full effect.

However, full implementation requires a great deal more work. For example, how should losses be evaluated? Who will evaluate losses? Are severe injuries caused by environmental pollution clearly distinguishable from those caused by other factors? What if the relevant administrative authorities do not transfer environmental criminal cases to judicial authorities? All these questions call for immediate attention as well as proper legislation and rulemaking.

C. *Broadening the Public's Role through the* Provisional Measures on Public Participation in Environmental Impact Assessments

Best practices reveal the importance of public participation in environmental matters. In some ways it is as important as administrative effectiveness. In spite of this, no well-developed mechanism for public participation in environmental protection had been articulated. Lack of access to environmental information undercut the public's ability to participate in public discourse. For this and other reasons, the public had virtually no role in environmental affairs.

SEPA issued the *Provisional Measures on Public Participation in Environmental Impact Assessments* (hereinafter referred to as the *Provisional Measures*) on February 22, 2006, to address these shortcomings. As China's first regulation specifying a role for public participation in public affairs, and China's first legal document regulating public participation in environmental protection, the *Provisional Measures* bears special significance. Details set forth in the guarantees public participation in the Environmental Impact Assessment process (EIA). The specifics of this participation can be summarized by a series of numbers: 5, 6, 7, and 10.

The number 5 refers to the five forms of public participation: public opinion surveys; expert statements; seminars; panel discussions; and public hearings that include experts. Public hearings have proven to be the most effective form of participation. The structure of the hearings has been worked out and largely standardized. They also have greater impact because of the face-to-face nature of interactions and the broader participation of more individuals. Hearings have their downside, however: they are complicated to organize and manage, and this tends to make them costly. Each kind of public participation has its advantages, and therefore should be used so that they produce the greatest overall impact.

The number 6 refers to the six situations in which public opinions should be solicited. They include the EIA for special projects as defined in Articles 11, 21, and 24 of the *Law on Environmental Impact Assessment*; projects requiring submission of an EIA report to authorities for approval; projects requiring reconsideration of an EIA; and other EIA activities.

The number 7 relates to the construction firms and environmental authorities responsible for publicly announcing environmental information regarding proposed construction projects and soliciting public opinion on related issues. These include bureaus and/or commissions formulating impact studies charged with publicly announcing five categories of information regarding construction projects within seven days of commissioning professional groups to execute EIAs for proposed projects. EIAs are then used to notify the public regarding projected environmental impacts and corresponding measures taken to prevent or reduce negative impacts.

The number 10 corresponds to the ten-day minimum that the public has for submitting comments on projects. Comments will be folded into applications for permits to proceed with project development and forwarded to relevant authorities for consideration.

The *Provisional Measures* requires that construction applicants consider numerous factors in choosing subjects for additional commentary in applications. Construction firms, government organs formulating impact studies, and professional EIA groups will solicit public opinions before submitting EIA documents for review and approval. Firms must represent different regions, occupations, and professional backgrounds among those offering opinions. A summary of these comments will go into the EIA, which must be made available to the public. Authorities will explain in the EIA report submitted whether or not to adopt the solicited public opinions.

The *Provisional Measures* has enabled the public to participate in environmental affairs by incorporating public participation into the EIA process. However, this limitation has blocked wider public participation in environmental protection efforts. Of particular note is the limited access the public has to information on environmental issues. Notably absent, as well, are provisions guaranteeing the public the right to seek relief for harm and damages in environmental cases. Given this, laws and regulations need to be drawn up and implemented that address these shortcomings. Environmental legislation experts and scholars from SEPA and China University of Political Science and Law have begun to address this very point by drafting the *Measures on Public Participation in Environmental Protection*. Hopefully, this decree will expand public participation in environmental affairs so as to fully realize and protect the public's access to environmental information, rights of participation, and rights to seek relief for environmental harms.

D. Regulations on the Importation and Exportation of Endangered Wild Animals and Plants: *Legal Guidelines on the Trade on Wild Fauna and Flora*

Under the auspices of the Legislative Affairs Office of the State Council, the *Regulations on the Importation and Exportation of Endangered Wild Animals and Plants* (hereinafter referred to as the *Regulations*) was circulated on April 29, 2006, and put into effect on September 1, 2006. The *Regulations* law was jointly sponsored by the State Forestry Administration, the Ministry of Agriculture, and the Endangered Species Import and Export Management Office. This law addressed both domestic demands and global pressures.

Illegal trade in wild animals and plants has become the world's third-largest illegal trade category, after drugs and weapons. Total trade volume in the category is estimated to be above ten billion USD.[4] In 1981, China became a party to the *Convention on International Trade in Endangered Species* (CITES). However, the CITES secretariat cited China's flawed implementation of the convention, noting that CITES principles have not been incorporated into domestic legislation, leaving many illegal trade activities uncovered by criminal laws.

China's remarkably rich and varied wild fauna and flora, as well as its relatively lax legal structure, has precipitated a rapid rise in the international trade of wild animals and plants. Total trade volume (legal and illegal) soared to 141.1 billion RMB[5] in 2005, up from thirteen billion RMB in 2002.[6] In the first half of 2006 alone, the trade volume

[4] "Annual Global Volume of Illegal Trade on Wild Fauna and Flora Exceeding 10 Billion USD," Xinhua News Agency (July 27, 2006) http://news3.xinhuanet.com/newscenter/2006-07/27/content_4885842.htm.

[5] Translator's note: The RMB (renminbi) trades at roughly 8:1 against the US dollar.

[6] Southern Jiangxi Forest Protection, "Speech Given by Jia Zhibang in the Training Program on Implementing *Regulations on the Importation and Exportation of Endangered*

topped four hundred million RMB, according to statistics given by Beijing Customs.[7] China has become a global leader in the production, consumption, and export of wild animals and plants. However, some regions in the country still see rampant illegal trade in these resources, posing a threat to biodiversity on national and global levels.

The *Regulations* was implemented in an effort to effectively protect and rationally utilize wildlife resources, to better implement CITES, and to better administer the import and export of endangered species. As the first administrative regulation in this field, it has set forth many detailed rules to support the mechanisms stipulated in such laws and regulations as the *Law on the Protection of Wild Animals* and the *Regulations on Wild Plants Protection.*

E. *The* Administrative Measures Regarding Pollution Caused by Electronic Information Products

The Information Technology (IT) industry makes use of a wide variety of materials, not all of them benign. Components include gold, silver, aluminum, copper, iron, petrochemicals, and plastics. Manufacturing and disposal create significant environmental pollution. For this reason, China must take responsibility for more effectively treating such wastes.

On February 28, 2006, the *Administrative Measures Regarding Pollution Caused by Electronic Information Products* (hereinafter referred to as the *Measures*) was jointly developed and circulated by the Ministry of the Information Industry, the National Development and Reform Commission, the Ministry of Commerce, the General Administration of Customs, the State Administration for Industry and Commerce, the General Administration of Quality Supervision, Inspection and Quarantine, and the State Environmental Protection Administration. The *Measures* took effect on March 1, 2007. According to the *Measures*, the prevention and control of pollution caused by discarded electronic information products forms a cornerstone of overall recycling and reuse efforts. Six

---

Wild Animals and Plants" (September 18, 2006), http://www.gnsb.net/news/shownews.asp?newsid=4739.

[7] Department of Wild Animal Protection, State Forestry Administration, "Identity Card Sought for 30,000 Wild Animals and Plants," http://www.cnwm.org.cn/wildlife/particular.asp?id=13529.

toxic or noxious substances are of particular concern in these efforts: four heavy metals—lead, mercury, cadmium, and chromium—and two flame retardants, Polybrominated biphenyls (PBB) and Polybrominated diphenyl ethers (PBDE). Monitoring of R&D, design, manufacturing, sale, and importation will allow regulators to track usage of these substances and to prevent them from entering the environment.

F. Provisional Regulations on Disciplinary Measures Regarding Dereliction of Duty in Environmental Protection: *Integrating Authorization, Supervision, and Protection*

It is widely known that laws and regulations currently on the books regarding disciplinary actions in cases of environmental dereliction of duty only outline general principles. The lack of specific guidelines and actions undermines efforts for prosecutors to take cases to trial. The *Provisional Regulations on Disciplinary Measures Regarding Dereliction of Duty in Environmental Protection* (hereinafter referred to as the *Provisional Regulations*), jointly issued by the Ministry of Supervision and SEPA on February 20, 2006, specifies for the first time the prosecutable acts of dereliction of duty, including among government officials charged with carrying out environmental policies and enforcement.

The *Provisional Regulations* was well received; they are brief yet comprehensive. Its sixteen articles reiterate all environmentally-related dereliction of duty offences as stipulated in existing laws and regulations. It also specifies the disciplinary actions to be taken for each offence. It includes language on the enforcement bodies to be brought to bear for various offences, and even outlines interdepartmental collaboration necessary to prepare and argue cases. The *Provisional Regulations* grants enforcement powers to officers, but still includes checks to prevent abuses of this authority. No other environmental law has integrated authorization of legal power with expanded monitoring powers, and tempered both with various checks to prevent abuses. This marks an import innovation in legal evolution regarding environmental protection.

The *Provisional Regulations* applies to state bureaucrats and employees, including law-enforcement officers and bureaucrats who may or may not work for environmental agencies, including government-appointed administrators of state-owed enterprises.

G. *Local Environmental Legislation: More Quantity, More Quality*

Numerous environmental laws on both the national and the local level were passed in 2006. In addition to being productive overall, legislators and regulators added a number of high-quality laws in localities.

Based on a study of the best practices in developed countries and the practical conditions in Shenzhen, the Standing Committee of the Shenzhen People's Congress worked out the *Regulations for the Shenzhen Special Economic Zone Regarding the Promotion of the Cradle-to-Cradle Economy*.[8] Ten basic mechanisms were put forward, including various metrics gauging progress in reducing waste and improving reuse and recycling. This law moves the province away from the single-minded focus on GDP growth and toward sustainable production and consumption.

On March 29, 2006, the Standing Committee of the Zhejiang People's Congress adopted the *Regulations of Zhejiang Province on the Prevention and Control of Solid Waste Pollution*. This precedent-setting local regulation requires that soil-contamination assessment tools be developed, and that effective restoration systems be deployed to treat affected areas. Further, the *Regulations* place the responsibility for cleanup on the shoulders of polluters responsible for causing each case of environmental harm.

The *Regulations on the Preservation of Urban Lakes in Nanchang*, the first local ad hoc legislation on the preservation of urban lakes, was passed by the Jiangxi Provincial People's Congress in April 2006, and was put into effect on May 1, 2006.

On July 13, 2006, the Zhejiang provincial government issued the *Measures of Zhejiang Province for the Supervision and Administration of Pollution*, which stipulate that environmental quality shall be an important factor in appraising local officials' job performance.

On August 16, 2006, the Standing Committee of the Shenzhen People's Congress issued the *Regulations of the Shenzhen Special Economic Region on Environmental Protection at Construction Sites*. Some of the highlights include:

1. Establishing reviews of environmental impact, especially for construction projects;

[8] Translator's note: English-language sources in China sometimes use the phrase "circular economy" to describe the processes and policies that connect design, production, and disposal of goods with the aim of reducing waste and increasing reuse and recycling. This process has been dubbed "cradle-to-cradle" economics in the West.

2. Creating mechanisms for prosecutors to take on litigation in environmental cases;
3. Stipulating prohibited and restricted industrial projects be reviewed and amended by environmental authorities as needed; and
4. Raising maximum fines for buildings violating environmental laws to one million RMB, which is the highest in the country.

The regulations cited above comprise only a small part of the total range of new laws, but this sample does touch upon some of the key features of these new laws. In particular, this new legal thrust focuses on a few key themes: localization, specificity and detail, and innovation in design.

II. *Revised Laws Reflect New Developments*

While these new laws were put on the books in 2006, older laws and regulations received substantial revisions. Some examples include the *Provisional Measures on the Management of Scenic Locations* (1985), the *Measures for Implementing a System of Water Well Licensing* (1993), and the *Measures on Environmental Letters and Petitions* (1997).

A. *Revision of the* Provisional Measures on the Management of Scenic Spots: *Integrating Preservation and Development*

In 2005, the crew of *The Promise*,[9] a film directed by Oscar-nominee Chen Kaige, was accused of causing environmental harm in the scenic setting of Shangri-La, Yunnan Province. A ninety thousand RMB fine was imposed in accordance with Article 27 of the *Regulations of Yunnan Province on the Administration of Scenic Spots*. The vice governor of Shangri-La County lost his job because of his dereliction of duty. A number of similar cases came to light following this, giving rise to a fierce debate on the need to protect natural reserves.

The *Provisional Measures on the Management of Scenic Spots*, issued in June 1985, is so outdated that it no longer meets current needs. For

[9] Translator's note: This film is also known as *Wu ji* in the Mandarin market and *Mo gik* in the Cantonese market.

example, the planning system for scenic spots is inadequate and the compilation and revision of the plan lags behind rapid changes. Current protection efforts have fallen short, often overtaken by rampant, and damaging, commercial developments. Organization and oversight remain in disarray. Setup of scenic spots is chaotic. Infractions have not been clearly defined and punishments remain vague. To address these problems, the State Council undertook a major revision of the *Regulations on Scenic Spots*. The Council circulated the revised law on September 19, 2006, and ratified it on December 1, 2006.

B. *The* Measures for Implementing a System of Water Well Licensing: *Comprehensive Regulation and Control*

Based on a major revision of the *Measures for Implementing a System of Water Well Licensing*, the State Council issued the *Regulations on the Administration of Water Well Licensing and Fees*, which complies with the basic principles of the *Water Law* and the *Administrative Licensing Law*. The *Regulations* streamlines the licensing procedure for pumping groundwater. It also incorporates the fee structure for water into state legislation. Accordingly, the state, as the owner of the water, can effectively regulate water resources with tools ranging from administrative management to financial incentives. Prudent application of these tools will enable the sustainable development of water resources and economic growth.

III. *Laws Serve as a Link between the Past and the Future*

While progress has been made, problems still persist. For example, most existing laws focus on environmental problems in piecemeal fashion instead of considering them as part of an integrated whole. Further, the current management structure for environmental protection is no longer effective under a market economy. This problem is aggravated by the low effectiveness of staff and management in these posts. Finally, the limited role for public participation continues to restrict many environmental issues to marginal status.

On November 28, 2005, SEPA issued the *Proposal for National Environmental Legislation during the Eleventh Five-Year Plan Period*. Pursuant to the *Proposal*, many laws and regulations were drafted and/or circulated in 2006. These include the *Regulations on the Administration of Water Well Licensing and Fees*, and the *Administrative Regulations on Pollution Caused by*

Construction in Marine Areas. Public input was being solicited regarding the *Regulations for the Safe Transportation of Radioactive Materials*. The *Measures for Environmental Pollution and Damage Evaluation* was put on the legislative agenda of the NPC Standing Committee. In the near future, SEPA is going to propose laws and regulations in seven major areas, including drafting the *Law on National Environmental Policy*, the *Law on the Safety of Genetically Modified Organisms*, the *Law on the Prevention and Control of Soil Pollution*, the *Regulations on the Protection of Biodiversity*, the *Ecological Protection Law*, the *Regulations on Monitoring Ecological Protection in China's Western Region*, the *Regulations on the Protection of the Rural Environment*, the *Regulations on the Prevention and Control of Domestic Animal Pollution*, the *Regulations on the Management of Ozone Depleting Substances*, and the *Regulations on Environmental Impact Assessment for Urban Master Plans*.

To ensure that these laws and regulations will be properly and forcefully applied, China is formulating a series of detailed rules describing the implementation of these laws. Examples include the *Regulations on the Prevention and Control of Environmental Pollution near Drinking Water Sources*, the *Regulations on the Recycling of Special Products and Packages*, the *Regulations on the Prevention and Control of Vehicle Emission Pollution*, the *Management Measures Regarding Follow-up Environmental Impact Assessment for Construction Projects*, the *Regulations on the Prevention and Control of Construction Noise*, the *Administrative Penalties on Noise Pollution*, and the *Regulations on the Prevention and Control of Noise from Social Activities*.

A. *Formulation of Principal Laws is Stressed*

China is currently drafting the *Energy Law* and the *Cradle-to-Cradle Economics Law*, both of which play important roles in the legal regime of China's environmental protection.

As a principal law in the environmental legal framework, the *Energy Law* will lay down the basic principles and institutions in the energy sector regarding the planning of consumption and distribution targets, exploitation of fossil resources, allocation of resources, distribution of different energy types across the economy, the development and use of substitutes for fossil fuels, and coordination of different operators and departments overseeing energy development, distribution, and use. Since the late 1970s, many administrative and local regulations, such as the *Coal Industry Law*, the *Electricity Law*, and the *Energy Conservation Law*, have been added to the books. At the national level, the NPC Standing Committee passed the *Renewable Energy Law*, in early 2005. However, a law addressing energy policy at a fundamental and comprehensive level has yet to be drafted. The encouraging news is that an inter-departmental team was set up by the State Council on January 24, 2006, to draft the *Energy Law*, which addresses this need directly.

The *Cradle-to-Cradle Economics Law* is crucial to China's overall economic reform efforts. Since the cradle-to-cradle economic concept is still relatively new to China, supporting macroeconomic policies, tax incentives, and technical infrastructure have yet to develop. These shortcomings will complicate and slow implementation of cradle-to-cradle polices. Therefore, the process can be helped along by drawing up and implementing laws that support these policies.

In its current form, the *Cradle-to-Cradle Economics Law* will create eight systems for realizing this fundamental shift in economic planning, production, and consumption. These eight systems include program planning, industrial management, development of technical infrastructure, demonstration of program effectiveness, and the development of producer responsibility into the consumption and disposal phase of the product lifecycle, which, in turn, includes green consumption, disposal, and recycling systems. The draft will be submitted for discussion by the NPC Standing Committee in 2007.

B. *The* Energy Conservation Law *is Being Revised*

Since the *Energy Conservation Law* enacted in 1993 fails to adequately address current conditions, the NPC Financial Committee set up a team to revise the *Energy Conservation Law* on March 23, 2006. It only took the team half a year to evaluate the existing law, examine its current application, and conduct studies on the necessity of executing a thorough revision. The revised law is supposed to be submitted for discussion at the NPC Standing Committee at the end of 2006 or in the first half of 2007.

C. *Legislation on Natural Reserves Moves Forward Amid Debates*

China is one of the richest countries on Earth in terms of ecosystem diversity and biodiversity. The establishment of reserves is one way to protect these natural resources. However, ensuring the effective operation of reserves will require an update of the *Regulations on Natural Reserves*, since this law focused on developing and protecting scenic spots, while largely overlooking rather than prioritizing ecological function and cultural resources. Therefore, many Chinese scholars have also suggested drafting a *Law on Natural Reserves*, which would fundamentally realign laws pertaining to natural reserves to reflect the growing understanding of the function of natural systems. To move the national conversation forward, the NPC Committee on Resources and the Environment solicited expert opinions on this subject in 2006. The Committee distilled these and submitted them to the relevant departments of the State Council, People's Congresses at the provincial level, and other relevant experts, organizations, and institutions for further comments. However, scholars and government officials have divided opinions on the proposed law. Some believe that the current, decentralized management model is not adequate for managing natural reserves, while many others hold that centralizing these functions through laws will prove exceedingly difficult.

In addition, SEPA is calling for a revision of the *Environmental Protection Law*, which was issued in 1989. The revision aims to hold local governments accountable for environmental conditions and addresses the problem of local governments' interference with environmental law enforcement.

APPENDIX 1

A SUMMARY OF MAJOR NATIONAL ENVIRONMENTAL LAWS ENACTED IN 2006

Issued by the Tenth National People's Congress (NPC):
Renewable Energy Law of the PRC
Animal Husbandry Law of the PRC

Issued by the State Council:
Regulations on the Importation and Exportation of Endangered Wild Animals and Plants
Administrative Regulations on Pollution Caused by Construction in Marine Areas
The Ordinance on Regulating the Yellow River's Volume
The Interpretation of the Law in Criminal Cases Involving Environmental Pollution

Issued by the Central Military Commission of the PRC:
The Chinese People's Liberation Army's Regulations Regarding Environmental Impact Assessment

Issued by the State Environment Protection Administration of the PRC (SEPA):
Provisional Measures on Public Participation in Environmental Impact Assessment
Regulation Regarding the Safe Handling of Pathogens in Laboratories
Regulations Regarding the Safe Handling of Radioactive Substances and Equipment

Jointly issued by SEPA and the Ministry of Supervision:
Provisional Regulations on Disciplinary Measures Regarding Dereliction of Duty in Environmental Protection by Government Officials

Jointly issued by SEPA, the Ministry of Information Industry, the NRDC, the Ministry of Commerce, the General Administration of Customs, the State Administration for Industry and Commerce, and the General Administration of Quality Supervision, Inspection, and Quarantine of the PRC:
Administrative Measures Regarding Pollution Caused by Electronic Information Products

APPENDIX 2

RELATED COMMENTS

I. Putting the Environment First Should be the Cornerstone of Law-making (Excerpt)

Recently, the State Council passed the *Decision on Implementing the Scientific Concept of Development in Strengthening Environmental Protection*. This policy document sets long-term guidelines for the country's environmental protection work. In particular, Article 8, Chapter 3 of the *Decision* clearly stipulates that "a balance must be achieved between economic growth and environmental protection." Furthermore, it introduces, for the first time, the concepts of "environment first," "protection first," and "no development zones" in certain regions.

Specifically, in economically developed regions with limited natural resources, environmental protection should be the top priority when drafting development plans. In ecologically fragile regions and key conservation areas, development programs shall be restricted. Suitable industrial developments should be introduced with prudence and restraint, keeping in mind the need to maintain healthy and balanced ecological function in all areas. In natural reserves and areas worthy of special protection, economic development programs shall be forbidden. Such a change of philosophy marks a strategic and historic shift in the guiding principles and policies of China's environmental protection work.

In terms of environmental legislation, we used to emphasize the balance between economic development and environmental protection. Now, we should give priority to environmental protection and guarantee it through detailed regulations and measures.

Wang Canfa, Director of the Environmental Resource Law Research and Service Center of the China University of Political Science and Law

Source: http://news.sina.com.cn/c/2006–0-2-20/08208250625s.shtml

II. Environmental Laws Are Mostly Ineffective
"We have quite a number of environmental laws, but most of them do not seem to work," said Pan Yue, Deputy Director of SEPA, in an interview on July 12, 2006. Yang Chaofei, Director of the Legal Office of SEPA, listed seven major problems in environmental legislation.

1. The current environmental legal regime has not covered such important areas as soil pollution, management of chemical products, biological safety, protection of biodiversity, and nuclear safety.
2. Some administrative measures have not been adapted to address current conditions, some lack a grounding in a broader legal structure, and still others call for continuous and costly follow-up.
3. Many supportive regulations, codes, standards, and policies needed to effectively execute and enforce environmental laws are absent, which hampers the implementation of specific laws and the broader legal structure.
4. No law regulates government behavior, which has allowed for rampant interference with law enforcement by local governments.
5. Civil compensation on environmental issues lacks a solid legal basis. Because of this, plaintiffs in such cases rarely receive adequate compensation for harm done to them.
6. An insurance system against environmental damage has yet to enjoy a sound legal basis. In the aftermath of major environmental disasters, no reliable funding can be quickly injected into affected areas. Consequently, either the government or the victims of such disasters have to shoulder the costs.
7. Punishments and penalties for environmental damage remain weak. In conditions where detection rates are relatively low, and enforcement costs remain high, the only way to raise the impact of environmental laws is to significantly raise the cost of violations. If the cost of breaking laws remains low, firms have little incentive to invest in preventive measures or to obey the law. In order to achieve results, legal and administrative changes must be made that will aid prosecutors and regulators in collecting evidence and enforcing court rulings.

According to Pan Yue, the current environmental legal regime must be reformed into a more centralized system. Key principles and all environmental laws and regulations should be incorporated into the General Provisions of the Constitution, including the need to develop environmental awareness throughout society. For example, the *Environmental Protection Law* should be revised to include specific programs designed to develop environmentalism.

Yang Chaofei also noted that the current *Environmental Protection Law* was formulated in 1979 by the NPC Standing Committee, and was revised in 1989 with no further revision for nearly twenty years. More and more people are calling for a second revision of the law. Representatives of the NPC and the National Committee of the Chinese People's Political Consultative Conference have repeatedly called for taking this step. According to Yang, conditions for undertaking a second revision of the law are favorable. More experts and policymakers want local governments to be responsible for environmental conditions. This notion must be integrated into new and revised environmental legislation so that local governments will no longer undermine environmental law enforcement.

Source: *Legal Daily*, July 13, 2006

CHAPTER SIX

ENVIRONMENTAL PROBLEMS IN DEVELOPING THE NEW SOCIALIST COUNTRYSIDE

Wang Peng

Environmental problems in rural China have become a matter of great concern. These issues have moved up in priority following the central government's development of an ambitious plan for "the construction of a new socialist countryside."[22] This plan, unveiled in early 2006, touches on agricultural, land management, environmental, social, and economic policies.

Problems endemic to rural areas, such as environmental deterioration, poverty, and overpopulation, hinder development. Farmers bear the brunt of these problems since they depend on the land and environment for their livelihood. This puts them at the center of tensions between economic development and environmental protection. Such tensions have long existed in rural China, and these conflicts are now intensifying.

Improving environmental protection efforts with the goal of developing healthy ecosystems in rural areas must be given a high priority in building a new socialist countryside. Doing so requires a thoroughgoing understanding of environmental conditions in rural China, especially the relationship between environmental deterioration and economic development. This knowledge can help to guide the successful development and implementation of environmental policies in China's rural areas.

Key words: new socialist countryside, dualistic structure, environmental problem, remediation

[22] Translator's note: A variety of related policies fall under this rubric. Poverty reduction plays a central role in the entire policy. Various steps have been taken to increase incomes and reduce production costs for farmers: direct subsidies and subsidized loans as well as tax relief have been two of the more popular and bold programs. Strengthening the social safety net has also lead to an improvement in services offered to rural residents, including improved access to medical care, primary and secondary education, and job training programs. Finally, the "new socialist countryside" seeks to repair decades (if not centuries) of environmental damage done throughout China. Reforestation efforts, water improvement campaigns, and other steps have been taken to help the land become more ecologically functional, and therefore more agriculturally productive. In essence, these programs draw upon the post-Revolutionary roots of Socialism by putting the social contract between the state and farmers at the center of policymaking and development.

I. *The Severity of Environmental Damage in Rural China*

Without effective environmental safeguards, rapid economic growth has caused serious environmental damage in China's vast rural areas. As documented in the media, people living in these areas, many of whom have fallen through the social safety net, are the ones most directly affected.

The causes of environmental problems in rural China are manifold. They are touched on below.

A. *Modern Lifestyle and Agricultural Industrialization*

As traditional customs have given way to modern lifestyles, the rural environment's natural balance has been upset. For instance, natural processes and traditional agricultural systems can no longer break down the volumes or types of waste generated in rural and exurban settings. Non-biodegradable products and packaging have proven especially problematic. Plastic bags and bottles litter rural areas, especially pesticide containers and various plastic products. Discarded electronic products and improperly discarded chemicals pose especially noxious hazards. These things accumulate for several reasons: lack of money needed to pay for their removal; lack of social and institutional infrastructure to motivate change; lack of management systems needed to coordinate refuse collection and processing; and lack of education regarding the proper treatment and disposal of refuse.

Various kinds of agricultural film and wrapping add to these problems, again because these products are non-biodegradable.

Biological waste also poses serious environmental and health threats. Animal waste, livestock and poultry killed by disease, and other germ and vector-carrying substances have become more common. Sewage, which rarely receives treatment in rural areas, has proliferated. Not only does sewage contain human and animal waste, it often contains various chemicals such as detergents and heavy metals, which can be especially harmful.

Clearly, any of these substances poses a threat. However, taken together, and in the massive volumes involved, the scale and complexity of that threat only multiples. The limited resources available for handling and treating these wastes further complicate matters.

Dumping sewage, chemicals, and carcasses into local water not only affects the people in a given locality, it sullies one of the most precious and necessary resources available to the entire ecological system.

Rivers and lakes have felt the brunt of careless dumping of waste. Direct poisoning, eutrophication[23] and the resultant algae blooms, and spread of contamination to groundwater and ocean areas have become increasingly common. In all of these cases, impacts are immediate and long-term. Not only do these forms of contamination limit access to clean drinking water, they also reduce the quality of water available for future agricultural use.

Water pollution in rural China is very serious. It has worsened both in intensity and in the total area affected. Typically, it begins as "point source" pollution, often originating with industrial or municipal sewage discharged from concentrated points, such as drainage pipes. "Area" or "non-point" pollution may start with point source pollution, but could be caused by other factors as well, such as acid rain. Because of this the origins of such pollution outbreaks, as well their full and lasting effects, are hard to pinpoint, prevent, and treat. As these impacts affect environmental conditions, they also affect agricultural production, farmers, and farming villages in profound ways.[24] China's preference for family-based, small-scale farming makes this type of pollution particularly difficult to control, because it can start from so many small farms; at the same time, educating so many small farmers on the impact of their actions has proven costly and difficult to coordinate.

Pollution, and its spread, is not limited to water. Air quality has become more of a problem. And like water pollution, the sheer volume and variety of waste has become a threat to air quality. For instance, burning waste can poison the air. Once airborne, chemicals can spread quickly over vast areas. With air and water quality affected, the overall quality of life suffers, since these two resources are the most important for maintaining human and ecological health.

Dumping wastes on open ground contains the impact in the short term, but because dumps are sometimes located near water sources, and because fumes do not restrict themselves to dumping grounds, water

---

[23] Translator's note: Eutrophication exists where chemical nutrients, especially nitrogen and/or phosphorus, increase in a given ecosystem. In aquatic systems it has serious consequences. Algae blooms often result, robbing water of oxygen, which threatens fish and aquatic animal populations. Volumes of nitrogen and phosphorus are so high in so many places that eutrophication has not only become a problem in rivers and lakes, but in oceans as well. In fact, entire sections of coastal waters have become virtual "dead zones" around river deltas.

[24] "A Sociological Analysis of Area Source Pollution in Rural China," http://www.chinado.cn/ReadNews.asp?NewsID=590 (accessed February 19, 2007).

and air quality are affected. Burying the waste offers better containment, but is not without its dangers, especially to soil and water quality.

Modern, industrialized agricultural production can be directly implicated in the proliferation of waste and the environmental harm it produces. Petroleum-based fertilizers, herbicides, and pesticides can foul land, water, and air alike. Granted, these things can increase crop yields, but excessive and indiscriminate use can contaminate soil, upset ecological balances, and produce health problems among people living in the agricultural areas.

Note that using these industrial products also disrupts tried-and-true connections between farming and culture. For instance, in traditional farming methods, farmers left cut down grain stalks in their fields to help enrich the soil and prevent soil loss. However, modern agriculture disrupts this sensible use of material because stalks are covered with chemicals and need to be removed after harvesting.

In 1962, Rachel Carson warned against the harm of heavy pesticide use in her seminal book *Silent Spring*. While her warning helped to bring about the elimination of DDT[25] use, industrial agriculture may be more dependent on chemical intervention than ever before. Notably, China has become the world's biggest user of fertilizers and pesticides. Usage has reached forty tons[26] per square kilometer, which easily exceeds the safe limit of 22.5 tons established by numerous developed countries to prevent soil and water contamination.[27]

B. *Urbanization and Local Industry*

Encroaching urbanization creates new environmental challenges for rural, exurban, and suburban areas. In the urban-rural fringe, where environmental planning and management are poor, mishandled waste and sewage presents a constant threat to the environment. Local indus-

[25] Translator's note: DDT, or Dichloro-Diphenyl-Trichloroethane, is one of the most infamous of synthetic pesticides because of the multiple environmental impacts it produces.

[26] Translator's note: Note that "tons" means "metric tons." One metric ton equals approximately 2204.6 pounds.

[27] Su Yang, "Guarding Against Environmental Pollution in the Process of Agricultural Modernization: A Problem that Can't be Ignored in the Construction of the New Socialist Countryside," http://www.chinado.cn/ReadNews.asp?NewsID=590 (accessed February 19, 2007).

trialization also contributes to environmental deterioration. Although township- and village-owned enterprises play an important role in rural economies, they have become primary pollution sources as well. Scattered across the countryside and lacking waste treatment facilities, these small-scale factories pollute vast areas as they discharge waste directly into nearby rivers and lakes. The impact of this carelessness can be seen in the massive fish kill in Baiyangdian Lake, Hebei Province, in March 2006. Untreated industrial waste from dozens of printing and dye factories and feather-processing plants left the lake, which happens to be one of northern China's largest natural lakes, lifeless.

C. *The Spread of Pollutants*

As the volume of pollutants increases, and the ways that they enter the environment proliferate, they affect ever-larger areas of land, water, and sky. Water-borne contaminants that originate and accumulate upstream make their way to areas downstream; waste produced in towns and cities along these rivers adds to the problem. At the same time, numerous products that end up becoming litter and waste in the countryside originate from cities and towns. Polluting industries in eastern cities advance westward into underdeveloped rural regions, and the increasing volume of waste generated in the West makes its way back into eastern cities in ever-increasing concentrations, typically by water and in the air.

Industrial, domestic, and agricultural sewage discharged into water upstream finds its way downstream into neighboring, and even distant, cities and provinces. The resulting pollution is severe, widespread, and difficult to control. The Tangbai River in central China's Henan Province, for instance, empties into the larger Hanjiang River in Hubei Province, a province that has suffered heavy losses because of waste from upstream areas. As upstream water sources empty into rivers and lakes, they concentrate in downstream water bodies, which increases damage and complicates clean-up efforts. Downstream regions feel the brunt of pollution's effects; upstream areas lack the resources or leadership to prevent these problems in the first place. Needless to say, this gives rise to conflicts between regions.

Pollution originating in urban areas spreads beyond cities. Large amounts of municipal wastes are buried, burned, or dumped just outside of cities. Spread by wind and water, urban pollution has had a disastrous impact on the countryside. In July 2006, *China Youth Daily*, a popular official newspaper, reported on the trouble suffered by villagers living near an open dump outside of Chongqing.

With industrial expansion, many companies and enterprises—including those creating grave environmental problems—are moving from eastern coastal areas to areas farther west. In remote villages, where supervision and control are not as strict as in developed areas, polluting industries find it easier to profit from and get away with environmental exploitation. For their part, local governments eager to attract investment often offer terms favorable to polluting enterprises. For the sake of revenue and prestige, some governments have even conspired with business leaders by turning a blind eye to environmental damage and its societal impacts. Such situations occasionally spark social unrest.[28] Almost all of the municipal governments in northern Jiangsu Province, for instance, bear responsibility for enticing firms to relocate in the area. Municipalities there have used their lax environmental policies and regulations to attract firms from southern Jiangsu Province, where environmental regulations are more restrictive. Since 1994, some 6,770 projects, each worth an average of more than five million RMB, have moved north. Among them are chemical, printing, and metallurgical firms.[29]

The city of Ji'an, in Jiangxi Province, serves as another example of government leaders targeting particular kinds of businesses to relocate in their jurisdictions. Note that as of 2005, about three-fifths of the city's forty-one paper factories had recently migrated from the neighboring coastal provinces of Zhejiang and Fujian.[30]

More outlying areas have been affected as well. For instance, the ecologicallysensitive environment in Inner Mongolia faces an increasingly threatening situation. Wastes produced by recently introduced enterprises pose an ongoing and worsening hazard to the grassland and the lives and health of local residents.

---

[28] Zhang Min and Guo Yuanming, "The Voice of Farmers in the Face of Pollution's Spread," http://news.xinhuanet.com/banyt/2006–06/30/content_4771824.htm (accessed February 19, 2007).

[29] Ye Wentian, "Polluting Industries Moving Northwards Amid a Wave of Investment Incentives," http://biz.zjol.com.cn/05biz/system/2006/08/09/www.zjsr.com (accessed February 19, 2007).

[30] Zhang Min and Guo Yuanming, "The Voice of Farmers in the Face of Pollution's Spread," http://news.xinhuanet.com/banyt/2006–06/30/content_4771824.htm (accessed February 19, 2007).

D. *Ecological Destruction*

Ecological disturbances produce local impacts as well as ripple effects beyond a given location. The impact on flora and fauna, including fish kills, species endangerment, deforestation, and land contamination are widely known. However, these localized effects can produce widespread impacts, including invasions from opportunistic species and outmigration of native species. Furthermore, primary disturbances, like soil and water poisoning, spread and produce secondary impacts, like species migration. In this way, the scope of a given disturbance can expand and take on new, unexpected characteristics. For instance, grasslands have become desertified. This process has taken hold for many reasons: long droughts, excessive grazing, and poor water management are chief among them. This, in turn, has threatened agricultural productivity as well as urban quality of life. Similarly, soil erosion and the degradation and destruction of farmland have produced complex and overlapping disturbances to the ecological balance in agricultural regions.

These effects may have unprecedented and unpredictable impacts on human populations. Based on current trends, some scholars project the possibility of large and widespread population displacement: they estimate that roughly 186 million people from twenty-two provinces across China—mainly in the country's vast western regions and ecologically-fragile areas strained by ever-growing population and resource mismanagement—will abandon their homelands. With few other options available, they will probably gravitate toward developed cities and provinces such as Guangdong, Beijing, Tianjin, Shanghai, Liaoning, Zhejiang, Fujian, Heilongjiang, and Hainan. These municipalities and provinces, however, can sustain no more than thirty million people, which means that more than one hundred and fifty million of the migrants will become "eco-refugees."[31]

Numerous factors have produced widespread and severe degradation of China's rural environment. Historical factors have played a role; Chinese have settled and worked this land for many centuries. Institutional factors mattered as well; small-plot farming sometimes drives farmers to push productivity to the limits of carrying capacity and beyond. Human negligence and carelessness have produced the

[31] Pan Yue and He Qinglian, "Systematic Obstacles to Environmental Protection in China," http://www.wyzxsx.com/2005/printpage.asp?ArticleID=12650 (accessed February 19, 2007).

most acute damage, though, especially since the economic development boom started in 1978. Pollution by numerous companies, illegal business operations, unauthorized mining and excavation, and destructive practices by farmers have all scarred and poisoned the land, water, and air.

"Solutions" for these problems come with catches as well as costs. Expanding tourism has been seen as a way of encouraging the growth of rural economies. Unfortunately, tourism, if incorrectly managed, can produce a new set a problems. For instance, litter left by tourists mars the land in many heavily visited locations. Profit-hungry firms that serve the tourist industry, like hotels, build and operate in a way at odds with maintaining the natural beauty and cultural resources they rely upon for their livelihood.

Other industrial uses have come with costs as well. Over-mining in the coal-rich Loess Plateau of Central China, for instance, has resulted in serious desertification, with significant loss of soil on which local farmers depend for their livelihood. On the other hand, overgrazing, especially by sheep, has left the grasslands in Ordos, the plateau region of Inner Mongolia well known for its fine wool sheep breeds, severely damaged.[32]

Even though outside forces have exacerbated the environmental problems at work in rural areas, local farmers have brought some of the problems on themselves. Those living in poverty-stricken areas, driven by hunger as well as greed, sometimes raid and overuse natural resources. For example, firewood is the main energy source for cooking and heating in rural China. When too many people harvest firewood, forests cannot regenerate themselves. As a result, desertification afflicts an increasing number of outlying areas.

To counter these problems, some local governments in northwestern China have made efforts to engage in environmental recovery since the late 1990s. However, most of their efforts have proven ineffective for systematic and technological reasons. More often than not, local governments and farmers have seen their efforts come to naught, as problems—once apparently solved—return with a vengeance.[33]

---

[32] Wen Tiejun, "A Few Words on Rural Environmental Protection," http://www.cef.ngo.cn/read.php?wid=347 (accessed February 19, 2007).

[33] Wang, Liwen, "An Analysis of the Unsuccessful Ecological Theory and Practice in Northwest China," http://www.ailong.com/pages/article/33/11.htm (accessed February 19, 2007).

II. *Causes of Environmental Deterioration in Rural China*

To solve environmental problems in rural areas, China needs to face all aspects of the current situation and address it from an integrated perspective. Doing so requires consideration of many factors: the role of farmers; the needs of agriculture; the complexity of rural ecosystems; the shortcomings of policy; and the force of economics. These conditions need to be thought of as interrelated. They need to be examined through a wide view that takes in foundational principles and wide-ranging implications. Only when we think about underlying causes and likely outcomes can we achieve real and lasting results.

A. *The Political Economy of Rural Policymaking*

Farmers are intimately involved with their land. However, in China farmers are relatively weak politically and economically. Therefore, they lack the ability to favorably influence their environment. In the absence of their influence, those with greater political and economic access and resources determine the nature, scale, and distribution of resources. Unsurprisingly, rural areas often come out on the short end of policymaking.

B. *Urban versus Rural Areas*

In China, a stark line of separation is drawn between cities and rural areas. This separation has played out in different ways over time. Before entering the process of reform and liberalization in 1978, administrative force dominated the policy process. Bureaucrats divided urban and rural areas by type of residence, employment, and education. Administrative dictums shaped organization and operations in rural areas.

According to sociology professor Sun Liping, the urban economy depended on the rural economy. However, after 1978, the urban economy lessened its dependency on the rural economy, while integrating more deeply into the global market. Now the two areas exist in different geographical and economic spaces. They also exist in two different institutional spaces: the government administrative system still holds sway in the countryside, while China's cities have embraced an increasingly market-oriented approach to organizing human activity and social decision making. "Overlapping dualism" has put rural and urban areas on fundamentally different paths. This structural split has had serious implications. Most infamously, urban and rural income and wealth have diverged sharply in recent years, and the gap continues to widen by most measures.[34]

This dualistic structure is also the underlying cause of many environmental problems in rural areas.

1. Basic social inequities have directly influenced the rural environment. For a long time, China has emphasized urban and industrial development at the cost of farmers' interests and the rural environment. Inequalities lay not only in the exploitation of resources that damaged the environment, but also in the migration of polluting industries from cites to rural areas, increasing the amount of pollutants being released into the countryside. The environment in rural areas suffered from development in both the countryside and cities.
2. Rural areas suffer from various resource shortages and mismanagement in implementing environmental policies. Capital, technology and equipment are in short supply. Public resources are not well supervised. Efforts to ameliorate environmental damage are ignored. Capital, technology, and human resources and their management almost always go to heavy industry, cites, and big projects rather than to rural environmental protection or sustainable agricultural production. In short, environmental work lacks the policies, organization, staff, and infrastructure to be effective.

   Urban environmental remedies that emphasize control of pollution while overlooking primary polluters leave rural areas to fend for themselves. In short, neither environmental policy nor its implementation is geared toward addressing the needs of rural areas.
3. According to social scientists Hong Dayong and Ma Fangxin, pollution in rural areas "to some extent, reproduces or strengthens the

[34] Sun Liping, "The Construction of a New Socialist Countryside through the Perspective of Urban and Rural Relations," *Business Weekly*, June 29, 2006.

current dualistic social structure, further widens the gap between cites and the countryside, and weakens the ability of rural areas to narrow the gap."[35] The vicious cycle of inequality between cities and rural areas has intensified environmental problems. Poor environmental quality hurts farmers' abilities to overcome poverty, and worsens social conflict and environmental crises.

C. *The Single-Minded Pursuit of Economic Growth, Globalization, and Consumerism Comes with High Environmental Costs*

Chinese society—including rural society—has been caught up in a frenzy of rapid economic growth. The obsessive pursuit of economic growth and material wealth at any cost has resulted in the unsustainable use of environmental resources. In rural areas where the economic, social, and political contract has been frayed, environmental costs have been especially evident.

The intrusion of modern culture has shattered traditional culture in rural areas. The allure of consumerism has infiltrated the lives and hearts of rural people. However, country folk have relatively limited access to the consumer culture's benefits. This has resulted in a poor man's consumerism taking the place of a farmer's traditional folkways. Traditional values are marginalized as less relevant but deeply seductive mainstream entertainment becomes more available. This led one commentator to note that "An underlying spiritual uneasiness and cultural anxiety has emerged in rural areas."[36]

The role of traditional culture in helping to maintain environmental balance seems to have disappeared. Global products have filled the vacuum left by the erosion of these cultural values. For example, internationally-known consumer products, such as Coca-Cola, can be found everywhere in rural areas. In a further cheapening of values, counterfeit copies of these goods have become equally ubiquitous in the countryside. Even as they gain access to poor copies of "The Real Thing," villagers typically face eroding access to general economic benefits and public services.

---

[35] Hong Dayong and Ma Fangxin, "A Sociological Analysis of Rural Pollution," *Social Studies* 4 (2004).

[36] Tan Tongxue, "Village Order, Cultural Reconstruction and Modernization Styles," *Dong Yue Comments* 2 (2006).

Consumerism, globalization, and growth-driven economics have atomized farmers' lives. They do not have access to modern or urban life, and at the same time the traditional structuresthey have long relied upon have been disrupted. Their way of life was never designed to cope with the vast demands of the contemporary marketplace. As a result, their social system has become more tenuous.

This is not just a battle of ideas; the detritus of consumerism litters the countryside as well as farmers' lives. The land they depend on, and the agriculture everyone requires, has also grown more vulnerable and more fragile. For instance, litter, in the form of plastic bottles, is everywhere. This is only the more visible sign of physical changes; worse problems lie below the surface, threatening the land and the lives dependent on it. Clearly, farmers and their agriculture depend on, and are vulnerable to, the physical environment as a community resource.

After China's entry into the WTO, the fate of Chinese farmers has been linked with the market economy and global trade system. In fact, China rapidly became one of the most liberal of all agricultural markets. Unfortunately, the long history of commodity markets and of agricultural economics in recent decades provides strong evidence that liberalization tends to drive down agricultural product prices. This has been the case in China in recent years. Unsurprisingly, farmers have faced an erosion of quality-of-life measures during this period.[37] This demonstrates that the global agricultural market is linked to rural social conditions, though most farmers remain unaware of this.

Changes that originate in other countries, far from China's countryside, can have a profound impact on farmers' lives. For instance, some developed countries have instituted strict environmental and food-quality standards on imported goods. These measures raise the costs of production for Chinese farmers, while imposing export restrictions on Chinese goods. Farmers unable to meet these environmental standards will be marginalized. Worse, farmers that meet standards may still become marginalized because they will be unfairly lumped in with those who cannot or do not raise their standards.

Farmers also face mounting competition from transnational businesses that have moved into the Chinese countryside and now compete with

[37] Tang Renwu, "Chinese Farmers under Globalization," *China Sociology and Anthropology Net*, Nov. 20, 2006.

Chinese farmers for resources and customers. This incursion further marginalizes Chinese farmers.

On the other hand, liberalization can open China to increased competition and entry from multinational companies. Some of these firms could enter Chinese rural areas and plunder resources there. At the extreme, Chinese agriculture would serve world needs before Chinese needs. This could worsen the crisis that Chinese farmers face.

Even though economic growth has blinded most Chinese to the needs of farmers, rural environmental problems must be addressed before they worsen.

III. *Dealing with Rural Environmental Issues*

A. *National Plan to Help Rural Areas*

On October 26, 2006, the State Environmental Protection Administration (SEPA) published the *National Rural Prosperity Environmental Action Plan*. The *Plan* takes preliminary steps in solving rural environmental problems between now and 2020. The overall goal is to bring environmental pollution in rural areas under control; to overcome "dirty, disorderly and damaging" conditions; to improve the rural life and work environment; and to strengthen the social contract regarding environmental security. Specifically, this final condition addresses the building of a new socialist countryside that includes "clean water, clean homes, and clean farms" as a way of improving farmers' living conditions.[38] This is an important measure taken by the central government to deal with rural environmental protection and restoration.

At present, some places are attempting different methods to solve rural environmental problems. Such positive explorations provide valuable practical experience in rural environmental protection and restoration. For example, in Shangba, a "cancer village" in Guangdong Province, a dam commissioned by provincial officials is under construction. This should help with water access problems. Two teams from Huanan South China Agricultural University and Guangdong Soil Research Institute already have settled in Shangba, dedicated to ecological restoration

[38] "SEPA Makes Plans to Solve Certain Rural Problems by 2010," http://www.xinhuanet.com.

involving plant life on Dabao Mountain. Professor Lin Chuxia from Huanan South China Agricultural University is exploring innovative forms of pollution treatment. If two thousand *mu* (six *mu* = one acre or one-fifth of a hectare) in Shangba are planted with plants that can be used as energy sources, the Dabao Mountain area will become the first testing ground for research on plants used as energy inputs in Guangdong Province.

B. *Localism in Environmental Policy*

Environmental protection in rural areas should take into consideration not only the situations of local farmers, but also the differing needs of farmers across regions.

Because China has a large population with limited land, it is impractical to attach too much importance to modern agriculture and urbanization. Environmental protection should include respect for Chinese rural cultural traditions. It should stimulate the inner vitality of rural society and establish a "new socialist countryside" with inner cultural vitality. Otherwise, the establishment of new socialist countryside would be just so much idle chatter.

Many experts on the issues related to agriculture, rural areas, and farmers call for a rural society that is, in both theory and practice, integrated through social, cultural, and economic development. Wen Tiejun, who works with the Rural Economy Research Center at the Chinese Ministry of Agriculture, has been a leading proponent of this view. He has founded the Yan Yangchu Rural Reconstruction Institute[39] to serve as an example of this approach's viability.

Non-urban areas are quite large in China. Farmers can be found in the plains, grasslands, alpine mountains, and minority regions. Environmental protection efforts must consider different geographical, topographical, and cultural differences between different rural regions. This more thoroughgoing approach will help to match development efforts into the particular ecological needs and constraints at work.

[39] Translator's note: Yan Yangchu (1893–1990), whose English name is James Yen, graduated from Yale University with a BA in Political Science and Economics in 1918 and obtained his MA from Princeton University in 1920. He devoted his life to educating illiterate persons and rural development. In 1943, he received the Copernican Award as one of the ten most influential "revolutionaries" at the time. Other recipients included Albert Einstein, Orville Wright, Walt Disney, and Henry Ford.

Furthermore, environmental polices must consider traditional culture. As Wen Tiejun said, the three features of new rural development should include three interlocking elements: sustainable agriculture; a harmonious[40] rural social system; and restoration of traditional culture and the natural environment.

C. *The Roles of Different Parties in Improving the Rural Environment*

Environmental resources are publicly owned. Therefore, managing them requires the active involvement of government, social institutions, and interested individuals, particularly farmers.

As a matter of fairness, the work of protecting the environment should be shouldered by everyone, not just by relatively poor rural residents. Therefore, rural environmental protection must receive support from the government, cities, and the more affluent. Beyond investment and other financial and policy supports, public participation is needed; an understanding of shared responsibility is necessary to power and sustain funding, policy changes, and advocacy efforts. All citizens should understand that the rural question does not stand in isolation. Rural pollution affects cities; the environmental, economic, and social issues at work in rural areas affect urban residents. Without question, the environment, with all its complex interdependencies, requires the attention of the whole society.

The small-scale farm is the building block of the rural economy, society, and environment. Lingering poverty, and the wide gap between living standards in rural and urban areas, must be addressed in order to achieve any measurable environmental restoration.

Generating substantive change will require input from farmers. Such changes need to produce tangible, visible improvements in their living standards in order to engage and inspire their wholehearted support for environmental causes. Policies that are radical, imposed from outside, or overly bureaucratic simply will not work.

To facilitate their participation, farmers should be encouraged to organize all sorts of associations. In fact, all manner of nongovernmental

[40] Translator's note: The term "harmonious" goes beyond connotations of calm, cooperative living. While these elements are part of the usage here, the notion of harmony brings together elements of Socialist and Confucian doctrine into an entire worldview that strives for commitment, coordination, and cooperation in households all the way to international engagements and policymaking.

organizations (NGOs) and public institutions should be encouraged to participate in rural environmental protection and rural development. These organizations have technological and educational resources as well as financial and social capital that can complement governmental efforts. They can help to develop rural environmental education and implement the use of environmentally-friendly technologies. They supplement financial supports available to farmers. NGOs and civil-society institutions are important forces that together can help rural society achieve environmental restoration and establish a new countryside.

D. *Introducing Environmental Policies*

Practically speaking, rural environmental protection needs the guidance of more effective environmental policy. Considering the realities of the agriculture situation, it seems doubtful that administrative environmental policies alone can resolve these problems. The standard operating procedures for dealing with these issues typically proceed like this: develop pollution-free, green and/or organic food; sort and recycle waste; establish green residences and developments, and so forth. These are important elements of larger programs, but they must be supplemented by other elements.

New approaches are needed. These new policies should be more realistic, taking into consideration rural economic and social conditions. They must be more comprehensive in addressing economic, social, and environmental dimensions. They must motivate farmer participation; without it, failure is assured.

An example of a new approach has been implemented by the environmental NGO Beijing Green Cross, which worked to establish a rural ecological district with local farmers in Wushan, Hubei Province. Their approach has been dubbed the "Wushan Model."[41] No doubt, the success of this model can be attributed to local policy considerations and inclusion of cultural elements. Nevertheless, the success of the "Wushan Model" is worth analyzing. Its general principles may benefit many other rural districts.[42]

---

[41] Translator's note: See Chapter 8, "The Wushan Model: Building a Sustainable New Socialist Countryside," in this volume.

[42] Translator's note: See Chapter 8 in this volume for a more detailed examination of the Wushan Model.

Globalization requires that Chinese agriculture products go on the international market. Green and safe products meet the demands of sustainable agricultural production and of worldwide consumer markets. Using market rules to change the thinking of farmers and to encourage agricultural research will promote development and the more widespread use of sustainable and environmentally-sensitive farm techniques and products.

While production is at the center of these changes, byproducts must be addressed as well. Clearly, agriculture produces large quantities of waste. However, with proper guidance, much of the waste can be used productively. This, in turn, can add to the economic base, creating jobs and income, while doing so within social, cultural, and environmental constraints. For instance, livestock and poultry wastes can be used as fertilizer, and slaughterhouses can make better use of new clean-production technologies. Reducing, reusing, and recycling byproducts would cut waste, damage, and inefficiency. In general, production would shift to low-impact systems.

APPENDIX 1

NEWS REPORT: "CANCER VILLAGES" FOUND IN MANY PLACES IN CHINA

More and more reports emerge each year regarding so-called "cancer villages." The growing number and visibility of these afflicted places points to the significance and scope of the problem. In most places in China, cancer could be caused by many factors: chemicals on or in food, poor diet, and exposure to various airborne or waterborne poisons. But cancer villages share something in common: they are located in heavily polluted places.

In many of these cases, high cancer rates can be blamed on unsafe drinking water. According to a survey conducted by the Ministry of Health and the Ministry of Water Resources, some three hundred million people in rural China do not have reliable access to safe drinking water. To satisfy immediate needs, many individuals drink contaminated water. For instance, about sixty-five million people regularly drink water with fluorine and arsenic concentrations far above permitted levels. Chronic arsenic poisoning, which occurs after long-term exposure through drinking polluted water, leads to skin and other kinds of cancer.

Xinhua News Agency, the state news agency of China, launched a series of undercover investigations in more than ten "cancer villages" along the Huaihe River in 2004. Huangmengying Village in Central China's Henan Province is the most infamous among them.[22]

Another village, Shangba, in South China's Guangdong Province, has seen over two hundred and fifty deaths from cancer since 1987. This is an extraordinary number given that on three thousand persons live in the village. Cancer is the official cause of death in more than 80 percent of the deaths there. The Hengshishui River, which flows through Shangba Village, was named one of the country's five most polluted rivers by a CCTV (China Central Television) program exploring economic issues in 2005.

---

[22] Ou Zhengtao, *A Secret Visit to the Huai River* (Beijing: Xinhua Publishing House, 2005).

The Tangbai River in Hubei Province is also in a sorry state, turning nearby Quwan Village into a cancer cluster and threatening the safety of drinking water for the one hundred thousand people along its banks. The river is polluted by domestic sewage and industrial discharge from midstream and upstream areas, where paper mills, alcohol distilleries, and fertilizer factories operate.

Other infamous "cancer villages" include Xiaojiadian Village, along the heavily polluted Dawen River in Shandong Province; Guangfeng Village, located close to a chemical industry zone in Jiangsu Province; and dozens of villages by the Le'an River in Jiangxi Province, where people have been dying from canc er since the 1980s. To compound problems, thousands of acres of cropland have been rendered infertile. From the far western regions of Xinjiang to the eastern coastal areas of Zhejiang Province, from the northern villages in Tianjin to the southern communities in Guangzhou, "cancer villages" arc found in many places.

Source: Various media reports

APPENDIX 2

RELATED INFORMATION: *A NATIONAL RURAL PROSPERITY ENVIRONMENTAL ACTION PLAN*

The comprehensive treatment of rural environmental pollution is a main goal of the *National Rural Prosperity Environmental Action Plan.* Under the *Plan*, organic waste is to be gathered in designated locations and processed in a timely fashion. It will then be recycled in appropriate ways, generally at county-level centers.

Sewage treatment is to be conducted to address local conditions and needs. The aim is to gradually introduce improvements like better home water management and the use of green plumbing technologies in order to control and cut sewage output.

Second, the *Plan* addresses industrial pollution treatment. Taking steps that address pollution should become a focus for every industrial firm, from plant design to day-to-day operations. In relatively developed eastern coastal areas, agricultural companies should promote clean production; in the central and western provinces, these firms should explore the use of natural alternatives to industrially produced inputs such as fertilizers and pesticides. Where possible, projects that produce heavy pollution should be banned.

Third, treatment of soil contamination will be part of the *Plan*. Based on national research findings, demonstration projects on soil contamination treatment and restoration should start in representative regions, according to different types of soil contamination, weather, topography, soil composition, and farming uses. Strict controls should be applied to prevent polluted water and effluent from finding its way onto farmland. Farmers should be taught the proper use of chemical products such as fertilizers, herbicides, pesticides, and agricultural film and wrappers. On the positive side, organic food production should be encouraged.

Fourth, the *Plan* will work to better protect and care for water resources. Environmental protection agencies should set up operations near key sources of drinking and irrigation water. These offices should be adequately staffed and otherwise maintained. Taking this step can improve water inspection efforts. Data needed to carry out water quality research could be gathered in these locations and sent on to labs qualified to complete chemical and other analyses. Oversight and protection efforts can form the basis of water protection districts. In general, the importance of water management needs to be elevated and maintained.

Fifth, the *Plan* will work to prevent large-scale pollution from animal wastes. To assure high standards of waste treatment, methods would include advanced animal waste treatment. However, to cut down on the volumes of wastes, wastes would be redirected toward gas production and organic fertilizer factories. Furthermore, preventive measures would play their part. Enhanced soil utilization and the use of safe and high-efficiency feed would cut down on waste and the volume of poisons found in wastes.

Source: State Environmental Protection Administration

CHAPTER SEVEN

RURAL SOCIETY COPING WITH POLLUTION

Tao Chuanjin

This research, produced by a team of field researchers from Beijing Normal University, draws upon survey data to provide an up-close and personalized account of industrial pollution in the Chinese countryside. The data uncover the kinds and sources of pollution in sixty-four villages, as well as the responses taken by villagers. The researchers found that preventing, controlling, and redressing the losses of pollution depends on a complex web of relationships among villagers, who often blunt intervention through feelings of loyalty, business operators, who double as employers, and government officials, who often side with business interests. The paper concludes with suggestions of ways in which villagers might increase their influence over local conditions, and possible roles for environmental NGOs in this process.

Key words: rural environment, pollution, public participation

Pollution caused by factories stands out as the primary environmental problem affecting rural districts. To better understand the problem students, from Beijing Normal University (BNU or Beishida) systematically documented cases of factory discharge in their rural hometowns during the 2006 summer vacation. In addition to monitoring pollution output, they worked to explore ways of responding to pollution incidence. This phase of the project drew on surveys and interviews of local residents. A total of sixty-four villages were selected as the subjects of this investigation. Table 7.1 summarizes these data.

Table 7.1: Geographical distribution of villages in the BNU pollution study

Provinces	Villages	Provinces	Villages	Provinces	Villages
Jiangxi	10	Xinjiang	3	Shanxi	1
Sichuan	8	Hebei	2	Zhejiang	1
Jilin	7	Heilongjiang	2	Jiangsu	1
Anhui	4	Henan	2	Tianjin	1
Chongqing	3	Shaanxi	2	Beijing	1
Shandong	3	Hunan	2	Guizhou	1
Hubei	3	Guangxi	2	Fujian	1
Yunan	3	Qinghai	1	Total	64

The sixty-four villages in the sample belonged to twenty-two provinces and municipalities. Northern and southern provinces were about equally represented. The western province of Sichuan, which has eight villages in the sample, is almost balanced by the eastern province of Jiangxi, which placed ten villages in the sample. The bulk of the villages lie in the midwestern region; outlying areas around major cities like Beijing, Tianjin, and Shanghai, and more developed provinces in the east, were underrepresented by design. Since many factories operate in these areas, the findings of this study may underreport pollution incidence.

The results show all types of factories causing environmental damage in and around these villages. Interestingly, the type of factory determined the degree of pollution, rather than any kind of regulatory intervention.

Rural dwellers had to rely on their own organized resistance efforts to ameliorate these conditions. As a result of their actions, they occasionally received some material or financial compensation for losses or damages suffered. However, the fundamental need to treat and prevent environmental harm has remained largely unaddressed. This may well be a function of the limited organizational, human capital, and financial capacities of locals, but certainly institutional and legal factors undermine such efforts as well. Given these limitations, alleviating pollution in the countryside demands more effective implementation and enforcement of laws by the central government. This will strengthen the hand of the locals and the environmental nongovernmental organizations (ENGOs) that advocate for the public interest.

I. *Industrial Pollution Remains Unchecked in Rural Areas*

This investigation reveals that a large number of rural factories are major polluters. Few of them used any form of exhaust or waste treatment, and therefore their discharges seldom met the regulatory standards.

Investigators classified pollution conditions three ways: light, medium, and heavy. Light pollution refers to a situation in which no factories existed in or around a village, or a village was not obviously affected by pollution from factories. Medium pollution prevailed when locals noticed pollution, but the environment was spared serious damage. Heavy pollution refers to conditions when the villagers believed that serious damage had taken place. The investigation shows that thirty-five villages were subject to light pollution; seventeen experienced medium pollution; and twelve suffered heavy pollution, as summarized in Figure 7.1.

Note that about one—in five villages suffers heavy pollution. Water pollution was the most prevalent form of contamination: paper manufacturers, chemical factories, and pesticide producers were the most common culprits. Specifically, paper manufacturers heavily polluted four villages and inflicted moderate pollution in three. One village moved back into the light pollution category after the paper mill there closed.

Firecracker and fireworks makers, alcohol distilleries, and oil drilling operations produced the bulk of heavy air pollution.

Table 7.2 and Figure 7.2 further illuminate these points.

Factories located in villages with light pollution stand apart. For the most part, these used relatively clean production processes. Furthermore, their location outside of central town areas and their water sources

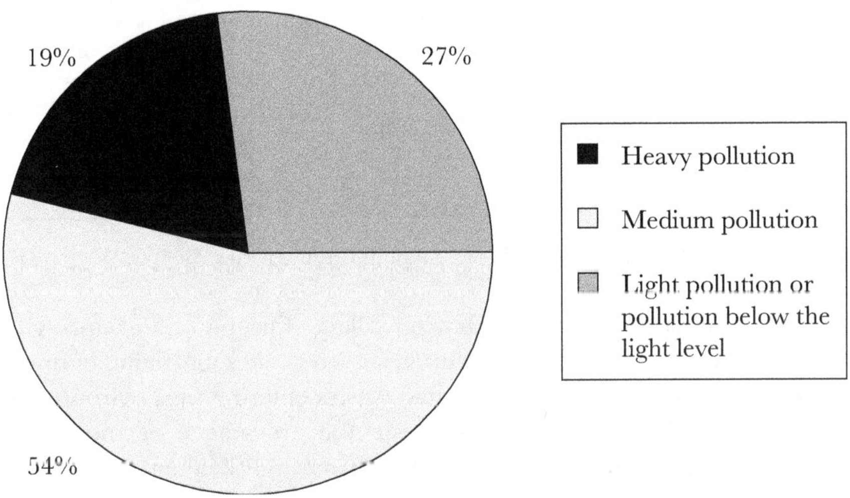

Figure 7.1: Degrees of pollution in investigated villages

Table 7.2: Types of pollution in villages

Medium pollution			Heavy pollution		
Water	Land	Air, noise and others	Water	Land	Air, noise and others
15	1	1	7	2	3

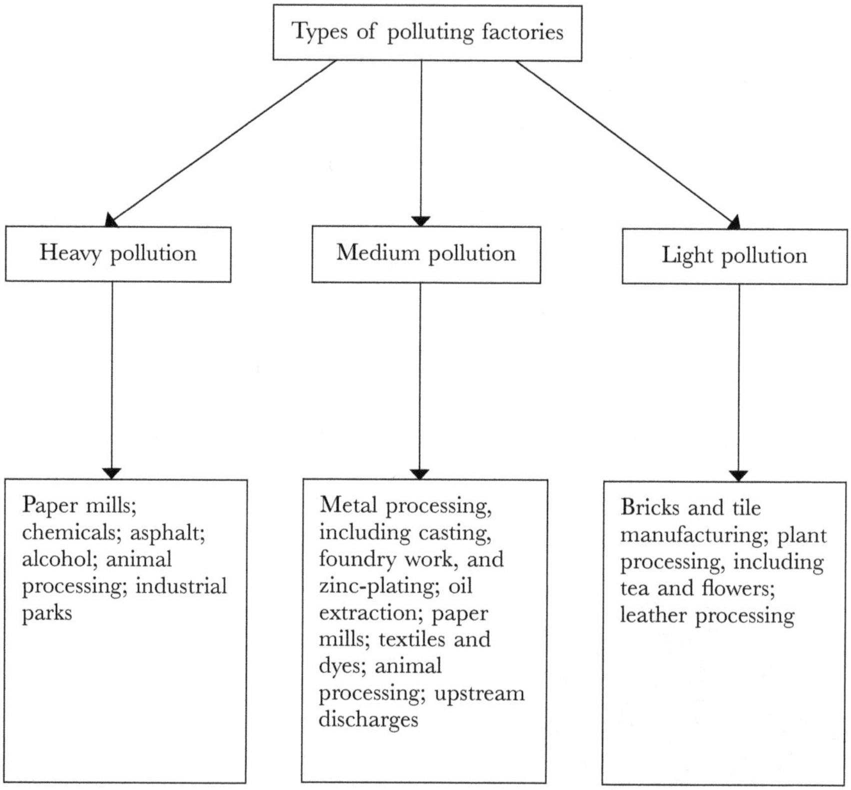

Figure 7.2: Types of polluting factories

mitigated impacts on the neighboring village. On the other hand, villages with heavy or moderate pollution served as sites for manufacturers using especially dirty processes, or had wastes entering the environment largely unchecked. Furthermore, given the prevalence of moderate and heavy pollution, it seems clear that exhaust and effluent treatment remains a low priority.

Table 7.3 provides descriptive details of the kinds of pollution involved, their sources, and their impacts. Note that the small and nonrandom sample used cannot be extrapolated to draw firm conclusions about the general pollution conditions in rural China. However, the findings are suggestive, and can be used to develop a larger and more generalizable data collection effort.

Table 7.3: Detailing the types, sources, and impacts of pollution in ten heavily-polluted villages

Types of factories	Pollution (as recorded by investigators)
Village 1: rock quarry	Smoke and dust from explosions adversely affected rice fields, harvests, and growers.
Village 2: paper mill	Six villages immediately downstream from the factory suffer the heaviest pollution. River water became foul-smelling, dirty, and bereft of fish. Several children who swam in the river on one occasion experienced itchy and spotted skin for a considerable time. Groundwater for drinking has become murky and yellow.
Village 3: paper mill and cement factories	Crop growth was badly affected.
Village 4: steel and biochemical factories	Children attending a primary school near the factories experienced the most acute adverse health effects.
Village 5: pig farm	Water in the village turned green and smelly, forcing villagers to fetch water from the mountains for daily use.
Village 6: pesticide factory	Seven villages experienced adverse effects, as well as a large stand of pine trees nearby, which formerly served as a substantial bird habitat. Despite reporting physical discomfort at work in the factory, many villagers continued working there for financial reasons.
Village 7: alcohol distillery	Wastewater from the factory had an unbearable smell. Worse, nearly all vegetation—trees, crops and other plants—around the wastewater outlet died.
Village 8: paper mill, oil refinery	A smelly, foamy effluent entered the river that flowed through four to five villages. Locals reported seeing dead fish floating in the river near this foamy substance.
Village 9: industrial park including a chemical factory, construction materials factory (gypsum, cement), and a paper mill	Many factories dumped wastewater directly into outside gutters, which blocked the gutters and caused them to back up. This standing water seeped into groundwater supplies; eventually well water in two villages became unsafe for drinking and irrigation.
Village 10: paper-making factory	Thirty years ago, the river running through the town was so clear that one could see its bed. Children used to swim in it. Now the river's water is black and white foam floats on its surface. The river emits a bad odor.

II. *Villagers Are Trapped in Polluted Places*

A. *Loopholes in Government Supervision*

It is common knowledge that governmental departments fail to effectively monitor and enforce environmental protection conditions and laws. The BNU investigation summarized in this paper supports this view. Villagers voiced numerous criticisms, making especially pointed comments about officials who take bribes. In general, villagers expressed dissatisfaction with government performance in dealing with industrial pollution in rural areas.

One investigator noted:

> I asked a relative of mine who worked in a factory that we investigated about government performance. He said local environmental officials were easy to bribe. When they raised issues with factory operators, things would be settled after they received meals or gifts.
>
> Other villagers told me that factories use no waste treatment technology. Factory operators wanted to save money; giving a few "red envelopes"[1] to officials cost less than installing waste treatment technology. The money kept officials' eyes closed to the existence and impact of the factory's pollution.
>
> Villagers had not reported the problems to higher government officials because they thought it would be of no use. They based their belief on experience. A few years ago, a big explosion at a firecracker factory killed and injured dozens of people working there. This drew attention from the central government for a time, but local officials hadn't done anything to improve working conditions since then.

Officials with governmental departments did not take their job responsibilities seriously. Furthermore, they did not seem to value environmental protection work. This led to lackadaisical work practices: "There were many reasons that the factories' emissions did not meet standards. A key cause is that local environmental supervisors never checked on the sources, types, or impacts of pollution."

Moreover, conflicts over jurisdiction and enforcement existed due to unclear divisions of authority. One investigator reported a case in which

[1] Translator's Note: The "red envelope" is used to enclose cash gifts. It can be used to give a cash gift in a formal situation like a wedding, but in this case it refers to a discreetly delivered bribe.

> Villagers reported the river pollution to the local government, but saw no results. Governmental departments there claimed that the district was on the border between two cities and was under the authority of neither.

In other cases, polluting factories took whatever steps were deemed necessary to escape punishment. One of the investigators provides a stark example of just how far operators would go to evade law enforcement:

> I was familiar with the owner of the paper-making factory because my family provided raw materials for it. He told me the factory had to spend millions of yuan on building a waste treatment pool and hundreds of thousands of yuan on a single bottle of chemicals to treat wastewater every one or two months. Besides the material expense, professionals had to be hired to carry out these procedures. Wastewater treatment would cost the enterprise about ten million yuan each year, which the operation could not afford. Therefore, the factory would treat wastewater when local officials came for inspections and go back to discharging wastewater directly into rivers after the officials left.

Evasion had become part of the businesses operations with other firms as well: "The factory had moved to this new location after years of production elsewhere simply to avoid government restrictions." Another investigator wrote that

> The paper-making factory treated wastewater in a special pit. Limestone was put into this pit. I went there and found the water was dark red with a bad smell. Because the factory was in a remote position across the river from the village, it had little impact on the villagers. About a dozen villagers worked in the factory. The pollution generated no public outcry.

B. *The Final Option: Public Resistance*

In some cases, ineffectual government supervision has forced villagers to take matters into their own hands. In order to protect their interests, villagers have organized publicly to resist the actions and negligence of factory operators. Thus far, this has been the most effective means to solve environmental problems.

One investigator wrote that

> Residents of two villages near a polluting factory went to a higher district government office to complain about pollution in their towns. A year later, the firms that had been named provided clean running water to most villagers free-of-charge. The industrial park where the factory operates has built a buffer zone around its perimeter and planted fir trees there.

Table 7.4: Village reactions to pollution

Reactions to pollution by residents in twenty-eight villages					
Employed some method of public resistance including a government role					
Yes				No	
11				17	
Results of resistance or assistance seeking				Reasons for not engaging in resistance or assistance seeking	
Compensation	Realization of Balanced Interests	Accomplishment of Zero Pollution	No Change	Tolerate pollution or leave	Economic concerns, such as job market impact
4	4	0	3	14	3

Such efforts do not necessarily eliminate pollution, but they can achieve measurable improvements in living conditions. Because of this, the BNU study found that organized resistance had become a popular means for addressing environmental problems. As in the case above, villagers often reported problems to local and superior government offices. They also resorted to public protests when other paths of recourse failed to produce change. Table 7.4 summarizes the forms of resistance used by villagers in seventeen moderately-polluted villages and eleven of the most heavily-polluted ones.[2]

Direct resistance returned somewhat unsatisfactory results. Although people sometimes received material compensation, the damage had been done. Note that in only four cases, or nearly one in three, villagers felt that they had received compensation commensurate with the losses they sustained. However, pollution was neither cleaned up nor curbed in these cases. Another four cases offered villagers only partial relief. In three cases no change took place. In no case did direct action result in the elimination of pollution.

[2] The villages with heavy and medium pollution were not separated due to the limited number of investigated subjects. One of them didn't reply.

III. *Protecting the Rural Environment from Harm*

Typically, villagers engaged in direct resistance only after their material interests had been directly and obviously damaged. No attempt was made to engage in preventive action. This indicates that villagers lacked a full understanding of industrial operations and environmental conditions, and the impact that the former could have on the latter. Worse, many villagers actively engaged in polluting activities, although many had no idea that wastewater had not been treated and had been dumped directly into surface water supplies. For the most part, though, both villagers and plant operators regarded economic interest above rural environmental and social conditions. Such thinking stifled some actions that might have otherwise produced positive change. This lack of awareness and poor opinion of environmental protection made it harder for villagers to coordinate and organize their actions.

Note that the impetus to take up public action has been damage to material interests, such as crops and cropland. The intrinsic value of the environment continues to be ignored. This privileges factory owners; if they bring sufficient economic benefit to a given locale, many villagers would support their efforts, as long as the villagers felt better off economically. Note the tension that this calculus sets into motion: some persons do well under these conditions, but by no means all. For instance, those who benefit from the factory, such as owners and workers, may want the factory to continue its work, even though it may produce harm to the general society or public interest. Similarly, the benefits enjoyed in the near term may produce harm in the long term. The point is that sustainability did not drive actions by factory operators or villagers. This ongoing problem undermines efforts designed to achieve true economic and social development in rural civil society.

Another example makes this tension more concrete. Villagers around casting plants and foundries did not clash with these firms because many residents depended on them for their income. As suggested in Table 7.4, many villagers turned a blind eye to pollution because their immediate economic interests were not affected. To combat such indifference—if not outright hostility—the environment needs champions to step in and advocate on its behalf. The narrow focus on near-term economic interest will not serve rural areas in the long term.

Social ties and a strong emphasis on showing loyalty also play a role in maintaining the status quo. A report from one investigator provides a clear example:

> A rich resident tending over 100 pigs enjoyed strong relationships with the other villagers. This individual dumped the pig excrement directly into gutters adjacent to his property. When asked about this, some villagers said that the befouled gutters seriously affected their lives and well-being, but they chose to accept these conditions, in spite of the resentment they felt. They did so because the pig farmer was on very good terms with most villagers.

Lack of faith in or experience with substantive change further undermined efforts to foment improvements in living conditions. Many villagers in the survey did not take direct political or social action because they doubted that such action would change anything. Little organization or coordination exists between individuals that could bind villagers together long enough to bring relief to problems. Instead, some families simply moved away from the most polluted areas.

For these reasons, public opposition to polluters, especially in the wake of government inaction, was a rare thing. By the time villagers took action, conditions had grown so bad, and the damage so extensive, that solutions seemed out of reach.

Even though the responses to pollution in villages leave much to be desired, some benefits can be seen. Perhaps most important, the public has become involved. Since villagers are the most directly affected, they need to play a stronger role in protecting the environment. However, taking this step should be made easier. Their actions should be protected and their efforts should be encouraged. They need to have a genuine role in providing input as to how to implement environmental policies. At the same time, the government role should be strengthened so that it can more effectively balance the demands of different interest groups while reducing pollution levels and their impacts in rural areas.

NGOs also have a role to play here. They can help to bring together rural residents and governments. The best and most committed NGOs see sustainability as a fundamental goal, on the local, regional, national, and global levels. Because of this, they speak for environmental interests. ENGOs can serve as educators, enhancing environmental awareness of rural residents and local government officials. This can protect villagers and help government workers to do their jobs more effectively.

ENGOs can draw on the best scientific knowledge available, as well as the most effective approaches for building cooperation and communication between different interest groups. By distilling reputable scientific findings into comprehensible forms, they can help villagers better understand the threats they face. This knowledge can also make it clear to government officials and plant operators that their negligence has serious ecological repercussions. Villagers armed with such information will make for a far more formidable force against polluters and negligent government officials.

By interceding among interest groups and bolstering the power of the largely disenfranchised villagers, ENGOs can help bring balance to the interest-group competition over rural pollution. As some interest groups increase thier power and organizational savvy, particularly industrialists, some countervailing force must emerge to protect the public interest. Leaving social decision making to those who can offer the richest bribes and those willing to accept them will fundamentally undermine the public interest, and will clearly disadvantage villagers.

ENGOs can strengthen the interests and arguments of villagers. However, because ENGOs do not have a direct, vested economic interest in taking sides, they can serve as neutral negotiators. The central government has partnered with ENGOs to organize and execute environmental policies at the grassroots level. Thus far, though, local and provincial governments have yet to work as closely with ENGOs.

CHAPTER EIGHT

THE WUSHAN MODEL: BUILDING A SUSTAINABLE NEW SOCIALIST COUNTRYSIDE[1]

Sun Jun

Beijing Green Cross (BGC), an environmental nongovernmental organization (ENGO) founded in 2003, has organized and implemented what they call the "Wushan Model" as a template for building the "new socialist countryside." Intensive fieldwork in Hubei Province has informed this model and helped to make it successful. This effort brings together sustainable economic development, ecological considerations, conservation, lifestyle changes, and spiritual pursuits to form a more meaningful and enjoyable living environment for peasants.

Key words: ecology-minded lifestyle, cultural life

Wushan Town lies within the jurisdiction of Gucheng County in Xiangfan City.[2] It is located in the mountainous area of northwest Hubei. In fact, it gets the name Wushan—meaning "five mountains" in Mandarin—from the five mountains within it: Yunwu Mountain, Ma'an Mountain, Qiujia Mountain, Bairi Mountain, and Lijia Mountain. It also lies on the banks of the Hanjiang River where it meets the Danjiang River. It is at the midpoint of the South-to-North Water Diversion Project (S2N).

Xiangfan City has a population of 5,800,000. Ninety-five percent of its residents live in the Hanjiang River Basin. In recent years, several sections of the Hanjiang River have become polluted. These areas have varied in width and length, but the overall impact has been devastating.

---

[1] Translator's note: Efforts to build a "new socialist countryside" include new approaches to development that reach beyond mere economic growth. Poverty reduction, tax relief, and subsidies have been the centerpiece of such efforts thus far, but they extend to education, cultural and social development, and institution building.

[2] Translator's note: Chinese jurisdictions and their naming may differ from some Western usages. For completeness, we list the administrative structure from top to bottom: Province, Prefecture, County, Town, and Village.

The river's water quality rating has eroded substantially. Most alarmingly, five of its tributaries are rated at Level V water quality.[3]

At the end of 2003, the Middle Line Project (the Hangjiang River section) of the S2N was launched. The project will divert approximately one-third of the perennial water volume of the Hangjiang to the water transfer system, where it will eventually feed into the Yellow River. This will substantially impact the river's carrying capacity for recovering from pollution, because both the total volume and the flow rates have been reduced. Although the project may ease water shortages in the north, it will impose serious difficulties on settlements in the Hanjiang River Basin. Helping locals cope with these new pressures and constraints will require them to learn new behaviors and production systems.

Since August 2003, Beijing Green Cross has conducted four field investigations in Xiangfan City. With the support of the local government and local environmental protection organizations, BGC chose Yanhe Village and Tianhe Village of Wushan Town as the pilot villages for building an ecologically sustainable countryside.

I. *An Introduction to the Wushan Model*

Yanhe Village, which lies about three kilometers west of Wushan Town, contains 225 families and a total population of nine hundred and ten persons. Its arable land consists of five hundred and fifty *mu*[4] of plowed fields and seven hundred mu of tea plantings. Villagers' 2004 annual per capita income equaled 2,980 yuan. Tianhe Village, which lies about five kilometers west of Wushan Town, contains five hundred and forty families adding up to a total of 2,219 people. Land holdings consist of 1,344 mu of plowed fields and over two thousand mu of tea plantings. Its 2004 annual per capita income equaled three thousand yuan.

---

[3] Translator's note: The Surface Water Quality Standard passed by SEPA in 1988 rates water in six categories. Level I is suitable for river source water in nature preserves. Level II is suitable for drinking water without treatment, swimming, Grade I Protected Areas, and rare and sensitive aquatic life. Level III is suitable for drinking water with some treatment, swimming, Grade II Protected Areas, and ordinary and unprotected aquatic life. Level IV is suitable for non-body contact recreational use and industrial use. Level V is suitable for agricultural and landscaping. Below Level V (Level V-) is completely toxic and unusable. Note that Grade I protected areas include river source areas and national-level nature preserves.

[4] Translator's note: One *mu* is slightly more than six acres or one-fifteenth of a hectare.

Beijing Green Cross focused its pilot study on three areas of reform: changing lifestyle and production methods; encouraging public participation in these changes; and adjusting local culture and beliefs.

A. *"Life First, Production Next"—Program of a New Lifestyle and Production Mode*

Daily routines and daily work are inextricably linked in peasants' lives. Therefore, lifestyle and behavioral changes can directly influence productive habits and resource management. This cuts to the heart of the "Lifestyle Changes Lead to Work Changes" campaign. This campaign aims to help peasants become more modern and well-mannered, and to use these behavioral changes as a springboard to influence the way they farm and use local resources.

Since its launch three years ago, the project has achieved significant impact. Both villages now have organized refuse collection and sewage treatment programs. Garbage and sewage are routinely sorted, recycled, and reused, when possible. Wastewater has been diverted from rivers.

BCG and other organizers have linked these changes to the need to make changes in domestic cleanliness and personal hygiene. Locals have learned how to more thoroughly clean their lavatories, bath areas, and kitchens. They now air out quilts and other material items that require it. They have also learned how to use dusting and cleaning cloths with greater care. These changes, in turn, have led to changes in personal care: villagers take care to keep their hands and feet clean, even clipping their nails and airing out their shoe insoles to reduce bacteria accumulations on their feet. While such actions may be taken for granted by many persons, these changes have a significant and direct impact on individual and family health, while at the same time protecting the health of water sources and soil in and around villages.

Collecting and sorting refuse has proven especially important, in that it steers peasants toward actions that promote cleanliness and community health. In short, this single program helps peasants get beyond some of their more coarse and crude thinking and behavior.

The garbage collection program was launched in March 2004. It is remarkably comprehensive. In both villages, households are required to have two garbage bins, one each for different kinds of refuse. Each village has a recycling center and several full-time sanitation supervisors who collect and classify garbage. "Wet" garbage is fed to chickens and pigs; feces are poured into methane-capture tanks;[5] and recyclable items, such as paper, plastic bags, plastic film and wrapping used in agriculture, and glass bottles are sold to the local recycling center. Bones and animal carcasses are used in compost. Hazardous wastes are moved into centralized storage; from there they are either incinerated or buried.

The step-by-step organization of the refuse collection and sorting system bears some explanation. In the case of Yanhe Village, the two garbage bins remain in front of each house. Households typically use plastic cans, iron cans, or small boxes. Twenty families even bought specially-designed yellow or green garbage bins. Sanitation supervisors collect garbage house-by-house, loading it onto trucks, which carry it to the local centralized processing center where sanitation supervisors further sort and process it. Tianhe Village does things a little differently: each household uses only one small garbage bin, and a single larger bin—usually a used oil drum—is placed every fifty meters in the village. The smaller bins are emptied into the larger ones. Sanitation supervisors then collect the garbage in these bins and carry the loads to a centralized processing center.

Parallel to all of this, each village made a landfill to collect otherwise unusable and non-recyclable items. Both landfills were dug using local labor.

Sanitation crews operate in both towns, but each town organizes them differently. In Yanhe Village, three teams exist, each including one team member chosen by villagers; Tianhe Village has four teams, each with four members. Sanitation supervisors, who are paid by the Village Committee, work at least once every two days. Their major responsibility is to collect garbage from household garbage bins, to transport this garbage to the processing center, and then to sort it for various uses and types of disposal.

To strengthen the educational aspect of these programs, schools introduced a "Green School" curriculum. The villages' two elementary schools have started environmental-protection educational programs. Among other program elements, pupils are discouraged from littering. Additionally, the Yanhe Elementary School pupils are required to keep an eye on their parents' work in sorting garbage. They also tour

[5] Translator's note: Many Chinese authors prefer the more archaic term "marsh gas" for methane.

the village every two or three weeks to check whether the garbage has been properly stowed in cans. Those who perform best in these tasks receive a "Little Environmental Guardian" award. Those who litter may be scolded by their parents.

Refuse collection and sorting has been a rousing success in Yanhe Village. The program has become part of the local routine. Villagers demand that refuse collection schedules be met. Even during planting and harvesting seasons, when many villagers were busy in the fields and refuse collection and processing was bogged down, villagers prodded town leaders, asking, "Where should I dump the garbage if you do not take the garbage away?"

Many urbanites treat refuse collection and sorting services as a given. Even in major cities like Beijing, Shanghai, and Guangzhou, however, not everyone has developed the habit of sorting and properly disposing of refuse. Too many persons fail to realize that carelessly discarded items could have been turned into useful items.

B. *Building Harmony in the New Socialist Countryside Requires Public Participation*

Socioeconomic development in Tianhe Village and Yanhe Village builds on similar work carried out in other rural development projects. A key element in this process is an effort called "Four-in-One Development." This venture includes four interrelated efforts aimed at improving sanitary conditions: building biogas tanks; improving water quality; bettering sanitary conditions in and around kitchens and lavatories; and maintaining tidier animal pens.[6]

In both villages, public participation has been a centerpiece for recruiting and successfully training members of households in these activities. In both villages, peasants are encouraged to participate in public affairs and solve practical problems collectively. For example, villagers traditionally harvested two thousand kilos of trees per household for heating and other energy needs. However, they realized that this volume cannot be sustained without causing lasting harm to nearby woodlands. Therefore, they discussed alternatives; from these collective discussions, the villagers developed plans to capture and use methane

[6] Translator's note: First launched in March 2004, the program has "graduated" 125 families in Tianhe Village and sixty in Yanhe Village.

from animal excrement. This change has allowed the forest to recover. Moreover, the liquid biogas' fertilizing effect on the soil allows farmers to reduce their demand for petroleum-based fertilizers and pesticides. This saves money and improves soil conditions over the long run. Organic tea farming thrives as a result. Locals have even attracted tourists to their Happy Farmhouse Project,[7] which demonstrates new, low-impact farming techniques and systems. Tourism focused around agriculture has opened up a whole new source of income and set of business activities in the villages. All of these innovations rest on a foundation of public participation and stronger civil society institutions.

C. *Village Code of Behaviors: Promoting Folk Culture*

Customs and beliefs shape behavior more than law. By the same token, material improvements alone cannot build a harmonious society in the countryside; the cultural and spiritual roots of a place need to be integrated into development efforts. This not only shows respect for local culture and people who make up these villages, but it also allows social innovations to build on well-established patterns rather than starting completely from scratch. A recent example from Tianhe Village underscores the benefits of working this way.

Villagers had originally planned to build a pavilion as a recreational and relaxation site. However, after extended discussions, villagers decided to draw on what they are known for, tea, to develop a "Tea Temple" at the village's edge. With the help of the GCB volunteers, they erected a building that has become a special cultural symbol of this village. The Tea Temple helps contribute to the diffusion of Chinese tea culture and consumption while generating unity among the villagers.

The Wushan Tea Temple officially opened on February 28, 2004, at a festive ceremony on a hillside in Tianhe Village. Villagers ritually prayed to Heaven and Earth and the God of Tea for a good harvest and a propitious year. These acts should not be looked upon as pure superstition; rather these acts express the sincere and pure joys and hopes of the villagers. Celebrations now include an annual King of the Tea Growers Contest as a centerpiece of the festivities. Taken together,

[7] Translator's note: A Happy Farmhouse is a rural household that has been granted a special license to accommodate and entertain tourists. This project increases farmers' income while allowing tourists to stay in a farmer's house and get a taste of the authentic rural lifestyle.

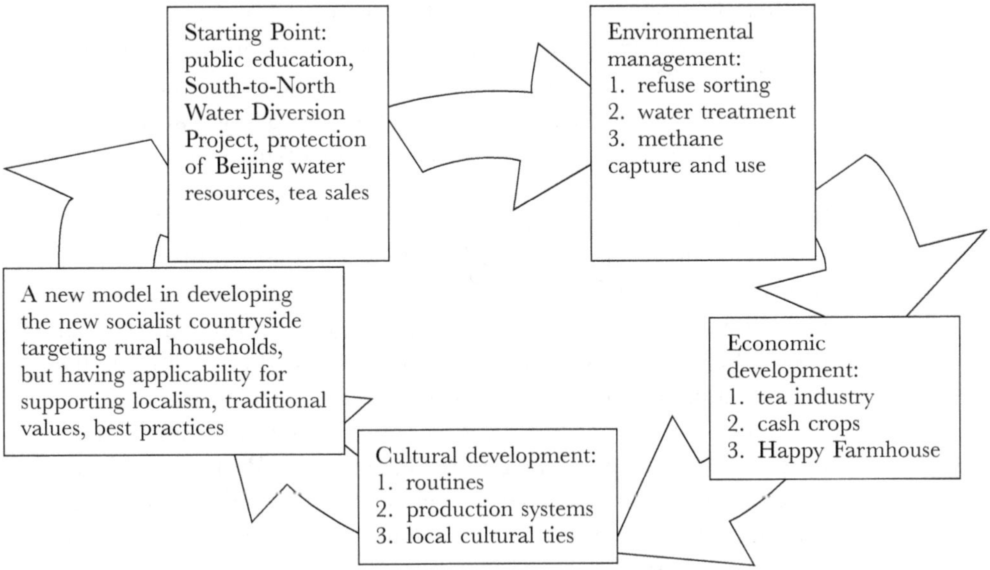

Figure 8.1: A graphic representation of the Wushan Model

these actions demonstrate the importance of linking cultural roots to public participation.

Villagers later founded a cultural exchange center based on tea growing and culture. This included a building for processing organic tea and tea tasting, displays describing the art of tea making, and other exhibits targeting tourists.

Facilities have also focused on providing better services for villagers. For instance, locals organized and opened a library. Water reclamation and reuse facilities have been established. Through such actions, lifestyle and culture work together with efforts to develop more effective livelihood and production systems. Unity is deepened and local culture enriched. Social innovations diffuse more widely and rapidly. The most backward and damaging behaviors in the villages have been eliminated.

II. *Analysis of Wushan Model and Summary of the Experience Gained*

As test sites, Yanhe and Tianhe Villages have gained many ecological, environmental and economic benefits from participating in this development project. Because it has worked so well, the model used in these two villages is now being applied to the other five villages of Wushan Town. To better understand the model and its application, a team from Beijing Normal University made a detailed assessment of the Wushan Model, especially noting the impact of the public participation components. Their assessment highlights the following points:

1. The project has lowered production costs and opened up new revenue sources for villagers. Linking methane-capture efforts to overall sanitation improvements, such as refuse collection and recycling and domestic and personal hygiene improvements, has brought savings and improved cleanliness. At the same time, the pig farming economy and culture has continued to thrive.
2. The intervention has improved economic conditions and opportunity. For example, the Happy Farmhouse Project in Yanhe Village opens up new revenue-generating possibilities. Through the Happy Farmhouse Project, villagers can make and sell new goods and services to tourists, such as local crops, entertainment, and lodging.
3. Environmental conditions have improved as environmental consciousness has been raised. Refuse collection and processing has greatly reduced plastic and metal litter. Increased use of organic fertilizers has lessened chemical pollution. Using methane for power has spared many trees in nearby forests. While each of these improvements is notable, the full impact of these changes taken together will not be seen for many years.
4. Social relations have improved, and neighborliness has become more common. Refuse collection and processing have established a clear division of labor. Accomplishing these and related tasks requires villagers to work together and take responsibility for the overall welfare of each settlement. The division of labor, which involves more men and children, has liberated many women from the drudgery of doing this work exclusively. Men and children have also become more involved in household cleaning chores. Many parts of the development work, such as building methane-capture pools, involved cooperation and communication among neighbors. Beyond this, larger projects, such as the construction of the Tea Temple, rely upon ongoing public engagement and communication.
5. Shared participation has raised the self-esteem of formerly marginalized individuals, such as the elderly and children. This has raised overall feelings of well-being among villagers.

6. Environmental protection has been given a higher priority throughout the countryside in Wushan Town. Take the example of how children have been affected by these programs. By granting the most dedicated and vigilant children the Little Environmental Guardians Award, children have been inspired to become more involved in environmental protection activities. For instance, they now participate in refuse collection and processing; these activities have become part of their daily routines. Their actions can then impact other family members, making each more keenly aware that every person bears some responsibility for protecting the environment.
7. This successful model will have a wider impact on society as it is replicated in other locations. The media have focused on refuse collection and processing efforts, as well as the move toward sustainable agricultural techniques. A press conference in Yanhe Village helped to launch a wider effort in building the new socialist countryside in Hubei Province. Soon after this event, the Peoples' Press released a five-volume series entitled *The Wushan Model* that details the work done there. Many governments at different levels have expressed genuine eagerness to learn from the Wushan Model and use it as the basis for establishing their own pilot village programs that unite economic, cultural, and ecological values.

III. *Further Reflections*

A number of key points have emerged through the Wushan test case. They are enumerated below.

A. *Human Power Is Unlimited; Funds Are Not*

The central government attaches great importance to the construction of a new socialist countryside. Unsurprisingly, the government has sunk substantial funds into such efforts. However, solely relying on central government funds would be a huge mistake, since its spending has limits. Moreover, not everything can be bought with money. In fact, many things sought through these programs have no monetary value: more refined lifestyles, ethical behavior, and good citizenship among them. We must also learn from past failures, especially with regard to urban development, which amply demonstrate that more money can exacerbate problems just as easily as it can solve them.

Values must play a central role in the social change process. The twin exemplars of Dazhai,[8] which showed us industriousness, and Daqing,[9] which demonstrated dedication, need to be recovered. After the 1960s, these examples have been considered increasingly quaint, but they still have much to offer. Public participation and cooperation need to be promoted as well. These are the two pillars upon which a solid foundation for change can be built.

B. *The Countryside Should Develop on Its Own Path*

Too many rural towns and villages have tried to imitate cities in their approach to development. This has produced many places that are neither truly urban nor rural; they exist in an uncomfortable limbo between the two types of development.

Because the countryside is different from urban areas, it should seek out and fulfill its own form of development. Thinkers, policymakers, observers, volunteers, and most of all residents involved in building the new socialist countryside should keep this aim firmly in mind. It will help to address needs more realistically while examining current conditions and long-term goals more honestly. Granted, this may slow down the rate of development; achieving goals may take twenty or twenty-five years rather than two or three, but this more reasonable pace will help to integrate local values and culture into new modes of behavior and production.

Village lifestyles should be respected and preserved. The history of each locality should be honored. The simplicity and genuineness of country living should be maintained and handed down to future generations. Villagers' fundamental industriousness should help to power change, but not at the sacrifice of their identity. By building on these values, we have greater hope in building a sustainable and meaningful new socialist countryside.

[8] Translator's note: Dazhai is the name of a village in Shanxi Province. It was regarded as the national agricultural model in the 1960s.

[9] Translator's note: Daqing, an oilfield in northern China, was held up as the national industrial model in the 1960s.

C. *New Leaders, New Peasants, and a New Countryside*

Building a new socialist countryside requires leadership. A large number of savvy, thoughtful, locally-respected leaders must be developed in order to sustain efforts to change rural areas.

To date, this has not been easy. Some grassroots leaders from outside a given village had no real connection to the land or the people where they worked. They did not understand market realities. They did not see environmental conditions in systemic detail. They did not match the industriousness and dedication of locals, nor could they keep pace with the cleverest and most creative peasants working in cities. These shortcomings must be addressed.

The overall quality of the leadership must be elevated. They need to better understand locals and local conditions. They need to develop a deeper understanding of agricultural economics at the village level. They need to understand environmental conditions as they stand, while seeing the impact of current activities on ecological conditions over time. They need to have respect for and understanding of local culture and history. They need to engage locals in productive public dialog and action. This new kind of leader can help to develop a new kind of peasant, and together they can build a new socialist countryside in China. Beijing Green Cross, through its efforts in Wushan Town, has shown one effective model for realizing these much-needed changes.

CHAPTER NINE

GREENING CHINA'S FILM INDUSTRY IN 2006

Guo Xiaojun

Film and television location shoots have produced environmental damage in various locations. Oft-cited problems surrounding filming of *The Promise*[1] highlighted these issues. The in the wake of the outcry over this incident, the public demanded the formulation of policies and regulations regarding outdoor filming, especially in ecologically fragile locations. Because of the entertainment industry's visibility, the push for "Green Filming" has become one of the most talked about environmental efforts in China. Advocates hope that greening the film and entertainment industry will help to deepen and broaden environmental awareness throughout Chinese society.

Key words: green filming, *The Promise* incident

I. *Damage to Bigu Lake by* The Promise's *Crew Triggers a Public Outcry*

Directors Chen Kaige and Zhang Jizhong, both household names in China, received nominations as Leading Green Figures in China for 2006.[2] The Association for the Promotion of Environmental Culture in China, with the co-sponsorship of seven government departments, grants this honor annually. The nomination came about in response to the shooting of *The Promise* at Bigu Lake in Shangri-La County in Diqing Prefecture, southwestern Yunnan Province. The film highlights the spectacular scenery of the area. However, the film's crew damaged the landscape and many persons believe that this demanded a response. By nominating Chen for the award, those unhappy with his role in these abuses, enumerated below, could very subversively call attention to his callousness.

The production team for the movie began shooting in the area in May 2004. The team invested over one million RMB in building three

[1] Translator's note: The film is known as *Wu ji* in Mandarin and *Mo Gik* in Cantonese.

[2] Translator's note: The reader should note, however, that any person can be nominated via Internet balloting. The list of nominees is posted online and left open for an online plebiscite. Winners are announced as part of a public ceremony.

work sheds near the lake: a so-called Crabapple Blossom Pavilion;[3] a simple wooden bridge leading to the island in the middle of the lake; and a 120-meter-long, 2.7-meter-wide trail linking the road extending into the woods to the lakeshore. These structures and modifications are spread over about five hundred square meters of meadow and brushwood.

Problems began to crop up in these structures after only a month, when welded joints in the pavilion began cracking. The local Quality and Safety Inspection Department issued a notice identifying the potential danger of this condition. After leaving the site in an emergency evacuation, the crew returned on June 14, when they signed an agreement with the Diqing government regarding management of the film location during the shoot. Among other things, this document stated that the crew would take responsibility for finding the pavilion construction crew and upgrading the facility. The agreement also made provisions for transferring all structures and "improvements" to the Diqing government at the conclusion of shooting.

Press reports claim that Chen selected the setting at Bigu Lake because it best fit his image of an undeveloped, natural landscape lush with flowers. When the flowers came into bloom, he called the location a "sea of azaleas."

On July 24, Yin Hongwei, a journalist with *Nanfengchuang Magazine*,[4] got a tip from tourists that the crew had damaged the landscape around the lake. He went to the lake to follow up on this lead. His report on the findings, "*The Promise* in Shangri-La" appeared in the magazine in August 2005.

The article begins with a description of the Crabapple Blossom Pavilion. Yin described the scene around the pavilion in vivid and shocking detail:

> On the side of the lakeshore which should be covered by green grass and wild flowers, there was a rough road created from sand, rocks and tree trunks. Near the side of the lake was a huge monstrosity constructed with concrete and wood in an as-yet-unfinished state. This was the famous Crab Apple Blossom Pavilion, over which the media had grown so agitated. In front of the Pavilion was a wooden bridge that was many meters long running across the lake, bisecting it. The bridge frame was based on iron

[3] Translator's note: The Chinese regard the crabapple as an elegant flowering tree found in special places.

[4] Translator's note: This means "south-facing window" in English.

> rods that had been plunged into the lakebed; wooden planks covered the bridge's walkway. It linked the Pavilion to the opposite bank of the lake. However, it was already bent out of shape in places; some wooden planks had broken loose. Anyone walking on the bridge risked falling off.
>
> Next to the lake, there was an open area of about four hundred square meters. Apart from the indications that some trees had been cut down, one can only say that it was garbage-strewn and hard to look at. Leftover signs marked "public restroom" and "workers only" lay on the ground. Abandoned house frames suggested that this place bustled with people not so long ago.[5]

As the film's visibility and record-setting box office performance grew, so did scrutiny of its environmental record. Numerous media outlets began a campaign of investigative reports, opinion pieces, and calls for reform.

This media coverage eventually caught the attention of government departments in charge of environmental issues. Deputy Minister of Construction Qiu Baoxing used his appearance at the International Symposium on the Control and Management of Water Resource Environments in Towns and Scenic Areas, held in Hangzhou on May 9, 2006, to fiercely criticize the film's crew.[6] On the following day, Zhu Xingxiang, Director of the Environmental Evaluation Department of the State Environmental Protection Administration (SEPA), said to the media that the Environmental Protection Bureau of Yunnan Province would be instructed to look into the incident and evaluate the severity of environmental damage caused by the film crew.[7] A week later, the Bureau released its report. The document enumerated various infractions: destruction of cuckoo habitat; damage to an alpine meadow; and littering.

With the degree of damage established, the nature of reparations had to be determined. According to the State Council's *The Regulations on the Protection and Management of Construction Projects in Wilderness Areas*, the production company would be required to restore the vegetation. Progress of their efforts would be regularly reported to public agencies.

---

[5] Yin Hongwei, "*The Promise* In Shangri-La," http://news.sina.com.cn/c/2005-08-17/15447523151.shtml (August 2005). An English translation of the full text can be found at http://zonaeuropa.com/culture/c20060129_1.htm.

[6] "Qiu Baoxing, Deputy Minister of the Ministry of Construction fiercely criticizes *The Promise*'s production team for its unacceptable behavior," (May 9, 2006), news.qianlong.com/28874/2006/05/10/2502@3161938.html.

[7] "The Yunnan Environmental Protection Bureau confirms damage caused by *The Promise* production crew," http://news.sohu.com/20060514/n243223541.shtml.

Various other punishments would be meted out as well, including penalties from the Diqing Environmental Protection Bureau.

The Ministry of Construction reported that on August 11, the crew was fined only 11,250 USD. Following this, a deputy county magistrate of Diqing was dismissed.

II. *A Legal Battle Gives Rise to a New Green Star*

Ke Lan, a former program host with Phoenix TV in Hong Kong, and the heroine of the movie *Legend of Shennong's Ladder*,[8] is a Leading Green Figures in China for 2006 awardee. However, as recently as July 2005, her frank and outspoken remarks regarding environmental protection, or the lack thereof, landed her in court.

At a July 2005 press conference during the Shanghai International Film Festival, Ke told the media that virgin forest in a Hubei national park had been damaged during a location shoot for the film. The Shanghai-based *The Xinmin Evening News* ran her comments under the headline, "The Production Team's Wrongdoing and I'm an Accomplice—Ke Lan Exposes Environmental Harm Caused during the Shooting of *Legend of Shennong's Ladder*." After reading the article, the films producer, The China Film Group Corporation, filed a liable suit against her and *The Xinmin Evening News* with the People's Court in Haidian District, Beijing.[9]

On July 17, 2006, the court reached its verdict in this dispute, addressing two related points related to the case. As to the nature and accuracy of her remarks, the Court determined that she simply stated her beliefs, which is allowed under the law. And since *The Xinmin Evening News* simply quoted Ke, the paper was determined to be blameless.

---

[8] Translator's note: The name *Shennong's Ladder*, or *Shennongjia* from the Mandarin, commemorates the legendary emperor, Shennong, who is believed to be the founder of herbal medicine and agriculture in China. The legend of the ladder is connected to his work as an herbalist: after hearing about special plants growing on a high precipice, he cut down a tall tree and climb it to reach these plants and add them to his medical collection. This wilderness is one of the most unspoiled locations in Hubei: old-growth forests are home to leopard, bear, wild boar and monkeys, including the endangered Golden Snub-Nosed Monkey. Legend has it that the Wild Man of Shennongjia, a Yeti—like creature, inhabits the area as well.

[9] "Ke Lan Taken to Court for Environmental Claim Regarding *Legend of Shennong's Ladder*," The Xinmin Evening News, http://ent.qq.com/a/20050619/000026.html.

As to the second part of the case, the Haidan People's Court determined that the production team working on *Legend of Shennong's Ladder* had entered the Dalongtan area of Shennongjia National Park without gaining permission from relevant departments with jurisdiction over park use. Furthermore, the wooden structures built on the set had adversely affected the environment. Both of these acts violated the *Regulations on Natural Reserves of the People's Republic of China.*

Upon receiving her Leading Green Figures award at a ceremony held on December 10, 2006, Ke Lan remarked that "I only did and said what I should...I don't think I deserve this award. But I do hope that no profit is sought at the expense of our children, whose sky should be clear and blue."[10] Ke then presented her trophy to a Mr. Hu, a scientific research fellow who has been working in Shennongjia for many years, to express her respect and support for his environmental protection efforts.

A similar incident occurred in early 2004. Li Jin, a singer from Abei Prefecture in southwestern Sichuan Province, submitted a complaint to SEPA alleging that environmental damage had taken place during the shooting of the film *The Legend of Condor Heroes.* As with the other films, this movie was shot in a fragile, remote, and protected area: Jiuzhaigou, a Chinese National Park and World Heritage Site in Abei Prefecture. Li claimed that illegally-erected sets had destroyed the vegetation at the Pearl Beach near Shenxian Lake. Other human traces had been left behind that would remain for years. He demanded that producers of the film issue a public apology, take remedial steps to correct the harm they had done, and compensate locals and the public for these losses. Furthermore, he demanded that producers commit to never engage in such actions again. However, his actions did not draw much media attention, probably because the film involved lacked personalities as well known as Chen Kaige and Ke Lan.

In fact, reports on the three aforementioned films only scratch the surface of a deeper problem. Numerous incidents of environmental damage caused by film crews came to light during 2006, thanks to some determined investigative reporting in the media and some well-known personalities speaking out.

---

[10] "Ke Lan: Giving Our Children the Clearest and Bluest Skies," http://www.ming-xing99.com/dalu/dalu/25537.html.

During August 2006, persons attached to the TV series *Clan Feuds* painted on the Xiandu Cliff Inscriptions in Jinyun County, a national cultural relics site in Zhejiang Province. This damaged the cliff's surface, especially its color. Cultural relic protection experts say that even remedial measures cannot restore the cliffs.[11]

The well-regarded magazine *Caijing* carried an article in its October 2006 issue entitled "An Impression of West Lake: Director Zhang Yimou Questioned on the Environment," in which angry local residents blame the director for breaking environmental laws during a performance called "Impressions of West Lake," which combined singing and dancing right on the lake.[12] A number of his movies, which include *Hero*, *The House of Flying Daggers*, and *Raise the Red Lantern*, have aroused environmental concerns.[13]

III. *Business Practices versus Environmental Protection*

What turned these well-regarded and famous directors into environmental villains? Two factors clearly help to explain the problem: large-budget films aimed at large global audiences, and a desire to include visually-stunning settings and breathtaking natural scenery. Without exception, these films feature grandiose scenery, boundless bamboo forests, tranquil lakes, azalea-covered hills, and many classic buildings.

While modest by American standards, the budgets for films like *Hero* and *The House of Flying Daggers* are huge blockbusters by Chinese standards: the former boasts a budget of thirty million USD, and the latter, thirty-six million USD. In the US market alone, *Hero* grossed over fifty-three million USD and *The House of Flying Daggers* pulled in over twelve million USD, making them two of the biggest foreign-language hits of 2002 and 2004. With a budget of over six million USD, Chen Kaige's *The Promise* raked in over nineteen million USD. *The Banquet*,

---

[11] "*Clan Feuds* Crew Destroyed Cliff Inscriptions in Zhejiang," http://travel.sohu.com/20060810/n244728015.shtml.

[12] "An Impression of West Lake: Director Zhang Yimou Questioned on the Environment," *Caijing*, www.dnzb.cn/gb/content/2006-10/30/content_2272626.html.

[13] Translator's note: Zhang seems to have reformed his behavior, however. In 2007, he and Feng Xiaoning won the Leading Green Figures of China Award. The award pointed to his efforts in hiring local peasants to restore grassy areas that had been damaged during the filming of a movie he directed. Feng won for making films with environmentalist themes. In some sense, these two awards underscore the very best of the "green filming" initiatives at work in China.

directed by Feng Xiaogang, took in twenty-two million USD on a budget of fifteen million USD.

These impressive numbers do not begin to account for the damage these films inflicted on the environment. Even with remediation, it will take years to heal the damage. Structures left behind, as well as the sites they stood on, have left deep impressions on the land. The oceans of flowers that attracted directors often suffered serious harm as a result of filming, and it remains to be seen when they will fully recover.

Liao Xiaoyi, president of Global Village of Beijing, a Sophie Prize[14] winner and a Leading Green Figure prize recipient in 2006, weighed in on film producers' penchants for shooting in wilderness settings and their irresponsibility in caring for the these precious lands, saying, "A wasteland in heart can lead to a wasteland in nature." She also pointed out that *The Promise* incident reflects a general lack of environmental awareness as well as a lack of ethical behavior in the entertainment industry. This is especially problematic because of the tremendous influence the entertainment industries have on the public. She expounded further on these points, saying, "On the surface this is a matter of aesthetics. On a deeper level, it has to do with our national spirit and values." She added that society has replaced nature with artificiality, casting aside the venerated traditional Chinese values of reverence for nature and seeking balance and harmony between man and nature. This shift in values has brought destruction and threatens both a shared spiritual legacy and our physical environs. These things reflect a general malaise in expressions of benevolence, depth of wisdom, and connection to beauty in the deepest sense of the word.

The fingerprints of local governments are all over these actions. For example, the producers of *The Promise* received a warm welcome and support from local governments in Diqing Prefecture. A working group headed by an official from the Diqing government was formed in Shangri-La County to assist with the shooting. "We are very happy to know that Chen Kaige chose to shoot at Bigu Lake," boasted one official from Shangri-La County to a journalist with *Nanfengchuang*

[14] Translator's note: The Sophie Prize is granted to persons and organizations that create "awareness of alternatives to modern-day development and/or initiated such alternatives in a pioneering or particularly inventive manner." Norwegian intellectual and author of *Sophie's World* Jostein Gaarder, with his wife Siri Dannevig, established the prize in 1997 through the Sophie Foundation. For more information, please visit: http://www.sophieprize.org/.

magazine. "It's a good opportunity to let the world see the beautiful scenery at Shangri-La and Diqing Prefecture."

Locations such as Shangri-La County in Yunnan Province, Zhangjiajie in Hunan Province and Jiuzhaigou in Sichuan Province, have become major tourist destinations. Local governments, and the people themselves, have come to depend on outside money. In order to capture money flowing into these places, economic development has taken precedence over natural conservation. Once this tendency sets in, and becomes part of expectations and decision making, it gathers momentum. In such conditions, it is no wonder that locals allow film crews on these lands. These productions add to local commerce during shooting, and serve to promote these locales after such films hit the big screen. It is more than slightly ironic that the film lingo of "green-lighting" a project can have so little green consciousness within it.

These conditions clearly show the tension between environmental protection and economic development in wilderness areas. People who live in these areas often find it difficult to reach a balance between the two extremes. Sadly, when things break down, they tend to favor economic development, sometimes in reckless forms. Locals sometimes violate environmental rules and regulations in their attempts to reap immediate economic returns.

IV. *The Film Industry Shows Support for "Green Filming"*

The Guangming Daily Press Group's coverage of these events in the *Beijing News* merits closer attention. In operation for only three years, the paper has become widely known among a diverse readership for its in-depth, professional, and responsible reporting.

The journal sprang into action only three days after Qiu Baoxing's critical remarks first hit the press on May 10, 2006, in an article titled "The Film and Television Industry in China: Have You Turned Green?" This article called for the industry to practice greater self-discipline and compliance with environmental rules and regulations. Furthermore, it called on the industry to promote greater environmental consciousness in films to encourage a general "greening" of society and the government, while supporting environmental protection organizations.[15]

---

[15] Qiu Baoxing, "Film & Television Industries in China: Have You Turned Green?" *The Beijing News*, May 10, 2006, http://yule.sohu.com/20060511/n243190686.html.

With the publication of this article, the tenor of the public discourse turned. Wang Zhongjun, President of Huayi Brothers & Taihe Film Investment Co. Ltd, commended the *Beijing News* for its "Green Filming" article and the attention it garnered. He stated his belief that the article is a warning shot across the bow of the industry regarding green issues. In Director He Qun's view, protecting the environment is intrinsically good. And Yang Yazhou, Director of the TV series *Nine Daughters in One Family*, said that the article taught him many things. For instance, he directed his crew to uproot green grass when shooting in Western China, since only yellow could be allowed as a background color. Now he looks back on his actions with regret and pity. "We can shoot more films, but we can't afford the environmental damage we leave behind," Yang said.

Li Xiaowan, producer of *The Door*, said

> Protecting the natural landscape and honoring the ecological balance while shooting is a rule we all accept. This has been proposed and emphasized ever since our first film generation.[16] Wherever you shoot, you must protect the local environment first and foremost.

Screenwriter Hai Yan once stated that in the past, film and TV crews were compared to locusts because they destroyed everything in their paths. Their nearsightedness undermined the public interest, and in fact the well being of their industry over the long term. Hai Yan added that, "Environmental protection is a virtue. But at present, it lacks both binding legal and moral structures. This leaves the environment vulnerable to commercial influences."

Wang Yongchen, a senior reporter for China Radio International and founder and head of the Green Earth Volunteers, a nongovernmental environmental protection organization, pointed out that one of the more valuable social functions of the film and TV industries is to positively influence public opinion. Greening the film industry serves this end because it facilitates greater harmony between human beings and nature. In fact, it arouses the public's interest in caring for and about Nature. This helps to counterbalance other tendencies that favor economic development at the expense of the environment.

[16] Translator's note: The Mainland film industry's directors are typically identified by the generation to which they belong. The first-generation directors worked in the early-twentieth century and established filmic practice and the Chinese film industry. This generation includes Zheng Zhengqiu and Zhang Shichuan.

Two months later, the First Summit of China's Film, Television and Tourist Industries was held on July 9, 2006, in Jiaozuo City, a film studio center in Henan. Under the auspices of the China Film Association, a "Green Proposal" was drafted by the Jiaozuo production center and thirty other participating entertainment businesses. The proposal appeals to industry workers to "care for nature" and to work toward"harmonious development" in the industry.

V. *Using Security Deposits and Other Innovations*

After word of the "Green Filming" proposal hit the press, cooperation between the film industry and environmental protection forces began. The security deposit aimed at environmental protection is an example.

According to press reports, on September 9, 2006, the production team of *Beidaihe, the Love Sea* signed an "environmental protection and responsibility contract" with the Beihaide government, the first of its kind in China. The production team also posted a deposit of 62,500 USD as an environmental bond with the government. Posting the bond, or deposit, operates something like an insurance contact between the production company and the local government.[17]

According to Tian Qiuyan, director of the Environmental Protection Bureau of Beidaihe, the production team checked in with Bureau representatives at the conclusion of shooting to make sure that both sides agreed on the wrap-up process. As part of this, the crew struck all temporary buildings and scenery. With Bureau staff in tow, they made sure that no environmental damage had occurred. Satisfied that the conditions of the contract had been met, the Beihaide government returned the deposit.

This process has now been standardized. Production crews interested in shooting in Beihaide, which lies northeast of Beijing where the Great Wall meets the sea, must now follow certain procedures. They have to first apply to the local government for access to locations and permission to shoot. Once they clear this hurdle, they must sign an "environmental protection and responsibility contract" with the local government and

[17] "Beidaihe: Natural Scene Undamaged, 'Environmental Deposit' Returned," www.he.xinhuanet.com/lvyou/2006-11/02/content_8413683.html.

post the bond of an agreed-upon amount. The local government will return the bond provided that producers meet the terms of the contract, as verified by local environmental protection departments. "Through communication and cooperation, we get to know production crews better. This also benefits the crew since we are familiar with the process and the land," said a Beidaihe government official.

Producer Zhang Jizhong's work in this area provides another enlightening example. On September 2006, he went to Yanguan Town in Haining City in southeast Zhejiang Province to shoot his TV series *The Story of Deer Mountain*. His reputation preceded him, since he received criticism for environmental damage associated with the shooting of *The Legend of Condor Heroes*. This time around, though, Zhang raised the bar for himself and his crew. At the shoot's opening ceremony, Zhang read a "Green Proposal" that declared: "It is our obligation to protect the shooting environment. We were ashamed of our earlier negligence." He signed a cultural relic protection responsibility contract with the local government, and put down a deposit of 37,500 USD. After concluding filming without incident, the local government returned the deposit.[18]

Parallel to this, government departments and bodies stepped up regulation of film and television shooting in wilderness areas. Article 13 of *The National Nature Reserve Supervision and Inspection Methods* (the *Methods*), implemented on December 1, 2006, states that

> Law enforcement and legal inspection in national nature reserves includes supervising and monitoring illegal tree harvesting, herb harvesting, grazing, hunting, fishing, land requisitioning, burning, mining, quarrying, earth removal, film and television shooting or other behaviors forbidden by the laws and regulations in the national nature reserves.

VI. *International Experiences*

Experts state that the Chinese film industry can learn from its counterparts abroad about how to be more environmentally aware. For instance, *The Day After Tomorrow*, a Hollywood disaster film, effectively communicates a message of environmental responsibility. While the film makes use of eye-popping computer-generated imagery (CGI) to

[18] "Zhang Jizhong paid a deposit of US$ 37,500," *Zhejiang Daily*, http://zjdaily.zjol.com.cn/jrzb/html/2006-09/27/content_1689513.html.

bring to life the consequences of environmental abuse, Roland Emmerich, the film's director, brought in a team to track the film's "carbon footprint."[19] To offset the carbon impact of the movie, the producers planted trees in California.[20]

Film commissions, industry groups, and watchdog organizations all play valuable roles in upholding environmental protection and animal rights. Film commissions ask for details of shoots and how producers will use locations. Environmental considerations enter into this process. Industry groups work to assure that their members uphold standards of good behavior. Watchdog groups, such as the American Humane Association, make sure that animals are not injured in the making of films. Films that meet their standards receive certification from the organization.[21] End-credit notices attest to producers' compliance with environmental and animal protection efforts.

When shooting *The Lord of the Rings* trilogy in his native New Zealand, Peter Jackson made sure that his crew replanted and restored vegetation that had been uprooted as part of the shooting process. To protect wild plants from being trampled on, they rolled out carpets wherever they shot, even in remote areas. The crew disassembled and removed sets and scenery at the conclusion of shooting.[22]

Now that the *Law on Environmental Impact Assessment* is in force, Chinese authorities and other capable parties can work together to produce environment and impact assessment before any film shoot. This helps to move the process away from neglect, where it started, past restoration, and toward prevention of environmental harm. Other laws and mechanisms now exist for treating damage when it does occur. These steps have made the film and television industries in China more green than they have ever been.

---

[19] Translator's note: An individual's or operation's "carbon footprint" measures the amount of carbon dioxide used in human activity. The footprint is used to calibrate the impact activities have on global warming. Various "offset" or cap-and-trade programs have been introduced as ways to reduce the impact of human activity on global warming. For more information: http://www.carbonfootprint.com/.

[20] "*The Day after Tomorrow*: Menaces from The Ice Age Brought About by The Greenhouse Effect," http://ent.sina.com.cn/m/f/2004-05-18/1157393037.html.

[21] Translator's note: Certification is noted in film end-credits, usually through one of two commonly used phrases: "No animal was harmed in the making of this film" or "Scenes appearing to place animals in jeopardy were simulated."

[22] "How Hollywood Protects the Environment While On the Set," *Nafang Daily*, http://www.nanfangdaily.com.cn/rwzk/20060521/zxgd/200606090050.asp.

APPENDIX

NEWS PLAYBACK: RISE OF ENVIRONMENTAL PROTECTION FILMS

Environmentally themed films rose to new prominence in 2006. This seems to dovetail with the zeitgeist, as reported by Xu Xin in the *People's Daily*, who wrote that "environmental protection" had become one of the major catchphrases of 2006. In her article, she explained that a profound understanding of environmental protection will lead to a more reasonable use of resources.

Numerous low-budget domestic releases plied this interest. With titles such as *Crazy Stone*, *You and Me*, and *The Forest Ranger* coming out in the winter and spring, and *Tokyo Judgment* and *Still Life* in distribution during the summer and fall, a steady stream of movies kept audiences thinking about these issues.

Set against the sweeping impacts of Three Gorges Dam, Jia Zhangke's *Still Life* depicts the lives of migrants displaced by the rising waters of the Three Gorges Reservoir. The film reflects the concerns of environmental and social activists alike. No two-dimensional polemic, the film won the Golden Lion at the 2006 Venice Film Festival. Not bad for a film made for just over one million USD.

Qi Jian's film *The Forest Ranger*[23] took home the Press Prize at the Shanghai International Film Festival on June 25, 2006. It was the only domestic film to win an award at the festival. Critics raved that "The film creates a bold and impressive image of a forest ranger, who is a respectable man struggling with his moral obligations and family demands, while facing powerful forces that oppress him and his true values."

The production team of *The Forest Ranger* fared well when shooting at the Wuyishan Nature Reserve. Deng Minli, vice director of the Office of the Wuyishan Nature Reserve, spoke to me on June 14, 2006, about the shoot. In this interview he noted that his office had closely

[23] Translator's note: Western readers should keep in mind that Chinese names are written last name first. Therefore, Jian Qi is the more typical ordering of this director's name in the West. The Mandarin title for this movie is *Tiangou*.

examined and finally approved the crew's request to enter parklands and to shoot within the park. All such effects had been aligned with the *Methods*. The crew and his office signed the required responsibility contract. Furthermore, the Office sent along a monitor who accompanied the crew throughout the shooting process.

"The crew planned to climb Huanggang Mountain to shoot some footage where the plants look like those farther north, but they finally gave up out of environmental concern," said Liang Shangyou, Director of the Management Bureau of the Reserve. "We are moved by what they did."

The controversial avant-garde director Zhang Guangtian released *Yuanmingyuan* in 2006 as well. This drama raises a question that haunts many environmental activists: "Who destroyed Yuanmingyuan?" This refers to one of the more visible and captivating environmental controversies in China during 2005, when construction crews attempted to complete an unapproved construction project in one of Beijing's most scenic parks. This event marks one of the first large scale public protests involving parkland.[24]

On August 3, 2006, the *Nanfang Daily* ran and article by Shi Yan entitled "What 'Yuanmingyuan' Shows is Environmental Protection." The article notes that support for the film comes from the China Environmental Culture Promotion Association, which also organized over a hundred college volunteers to participate in more than twenty performances in conjunction with the film's release.

Zhang believes that his way of dealing with environmental protection is more of a cultural approach. He notes the distinction between his work and that of environmental NGOs, pointing out the theme of his work is quite simple: if a wasteland appears in the heart, an oasis will never appear before the eyes.

---

[24] Translator's note: A blow-by-blow examination of this conflict is detailed in the article "The Yuanmingyuan Park Lake Incident and Its Impact on Public Awareness" by Zhao Yongxin. This article appears in *The China Environment Yearbook (2005)* (Leiden: Brill, 2007).

CHAPTER TEN

THE EVOLUTION OF INTERNATIONAL NGOS IN CHINA: BROADENING ENVIRONMENTAL COLLABORATION AND SHIFTING PRIORITIES

W. Chad Futrell

China's environmental issues now receive an extraordinary amount of international and national media attention. While sandstorms originating in China's north and northwest buffet Korea and Japan, coal-fired power plants pollute the air, literally adding fuel to the fire that drives global climate change. Governments and international non-government organizations (INGOs) are now working with their Chinese counterparts to mitigate desertification and climate change and to protect some of the most biologically diverse areas on Earth. From environmental education campaigns to eco-tours, from model cities to integrated conservation and development projects, INGOs have been at the forefront of environmental cooperation with China. INGOs spend over two hundred million USD a year working to protect China's environment.

In order to understand the evolution of INGOs in environmental protection in China, it is necessary to understand the larger political-economic contexts in which they operate. Indeed, shifting political winds often decided whether INGOs could operate at all within China's borders, not to mention what they could do. Furthermore, China's engagement with NGOs has always been part of a broader engagement with other international organizations working within the realm of environmental protection. For example, the Stockholm Conference and Rio Conference both mark definitive shifts in how the Chinese government viewed international cooperation around environmental protection. Finally, INGOs have interacted with a growing number of domestic actors, from national, provincial, and local environmental protection and forestry bureaus to the growing number of domestic environmental NGOs (ENGOs). The World Wide Fund for Nature (WWF), for example, worked with over two hundred partners on over forty projects in 2006 alone. Throughout their time in China, INGOs have both shaped and been shaped by how domestic institutions conceptualize and implement environmental protection measures.

I. *Introduction*

As Keck and Sikkink noted in their seminal work on transnational advocacy,[1] non-government actors were active in international politics long before the social sciences were codified, before scholars and government officials began using terms such as "non-government organization," "non-state actor," or "non-profit organization," or related concepts such as civil society and social movements. Each of these terms has its own intellectual history, including its application to environmental politics and Chinese contexts. For the purposes of this paper, I employ Salamon's five-point definition of NGOs as being groups that are private, organized, non-profit distributing, self-governing, and voluntary.[2] I have chosen this approach in part because it is often used by Chinese academics, including the purposefully-named NGO Research Center at Tsinghua University.[3] Along with Salamon's criteria, it is important to note that NGOs are often driven by principles such as human and women's rights, peace, and environmentalism. Here I will discuss the evolution and roles of INGOs in China, especially in regard to environmental protection. As will be seen, some of these INGOs are not environmental NGOs, per se, but their large roles in promoting environmental science, law, policies, and awareness nonetheless merits discussion.

The number of INGOs working in China at any given point is extraordinarily difficult to know. This is due in part to the fact that the Chinese government has not rigorously documented either domestic or international NGOs. Indeed, academics have only recently begun offering classification schemes for analyzing the emergent third sector.[4] Furthermore, many INGOs have been active in China for some time, but have yet to establish offices. Other INGOs have run specific programs in China for limited amounts of time, but have abandoned this work when funding ran out. Given the difficulty of providing concrete data on organizational types and numbers, this paper delineates some

---

[1] Margaret E. Keck and Kathryn Sikkink, *Activists Beyond Borders* (Ithaca: Cornell University Press, 1998).

[2] Lester Salamon and Helmut K. Anheier, *Defining the Nonprofit Sector* (Manchester: Manchester University Press, 1997).

[3] Wang Ming, ed., *Zhongguo NGO Yanjiu* (*Research on NGOs in China*) (Beijing: UNCRD Research Report Series, 2000.)

[4] For an example, see Ma Qiusha, "Defining Chinese Nongovernmental Organizations," *Voluntas: International Journal of Voluntary and Nonprofit Organizations* 13.2 (2002).

of the more dominant actors, tracing their activities in China in order to gain a better understanding of the broader processes in play and how they have affected environmental awareness and protection in China.

II. *Missionaries, Medicine, and Environmental Health*

Although environmental NGOs such as the Society for the Protection of Animals and the Sierra Club started as early as 1824 and 1851, the first INGOs in China were driven not so much by environmental principles but by religious ones. Christian missionaries had been coming to China for centuries, but their number increased dramatically in the 1800s. Some of these missionaries sought to spread Western science and medicine in China along with Christianity. The Medical Missionary Society began working in Guangzhou, Guangdong, and Macao in 1838, later working in other treaty ports.[5] The China Medical Missionary Society (CMMS) formed in Shanghai in 1886, continuing the Society's mission of spreading Western medical science and informing the West about China, in part through its *China Medical Journal*. Though these groups were not concerned with nature, they were some of the very first to be concerned with the environmental health issues that were increasingly affecting China as it developed economically.

Indeed, economic expansion into China's frontier areas in the west and southwest exposed new populations to animals carrying the bubonic plague. This expansion of trade networks thus led to increasingly severe outbreaks of the plague, which then traveled back eastwards along the same trade routes.[6] The plague claimed even more lives as it moved towards the heavily populated areas of southeastern China, including Hong Kong. The Pasteur Institute, a scientific NGO established in France in 1887, responded to the epidemic by sending the scientist Alexandre Yersin to Hong Kong in 1894. Yersin discovered the plague's pathogen, *Yersinia pestis*, which was subsequently named after him. A Pasteur Institute outpost was later established in Chengdu in 1911,

[5] Information on medical missionaries in China can be found at the University of San Francisco's Ricci 21st Century Roundtable on the History of Christianity in China: http://ricci.rt.usfca.edu.

[6] Carol Benedict, *Bubonic Plague in Nineteenth-Century China* (Stanford: Stanford University Press, 1996).

which was active until 1927, after which another Pasteur Institute office operated in Shanghai from 1938 to 1950.[7]

The Shanghai chapter of the Young Men's Christian Association (YMCA) was established in 1900, while several Protestant missions joined together to found the Union Medical College in Beijing in 1906. The Rockefeller Foundation came to China a few years later, purchasing and renaming the Peking Union Medical College from the missionaries in 1915. The Rockefeller Foundation shared some of the same principles as the medical missionaries, especially the promotion of Western medicine and science, investing a total of thirty-seven million USD in China from 1913 to 1933.[8] The spillover from Western medicine to other branches of science can be seen by the formation of the Science Terminology Association in 1917. The first congress of the CMMS had established a Terminology Committee in 1890 in order to standardize the translation of medical terms. This Committee became the Medical Terminology Association in 1915 before changing its name in 1917 in order to reflect the broadening interests of its members. Though these medical missionaries, doctors, and scientists were not working to specifically protect nature, the standardization and translation of scientific terminology fostered international exchanges among natural scientists and introduced many concepts to China. Indeed, there was no word or accepted translation for the Western conception of nature until '*ziran*' was adopted in the early twentieth century.[9]

The initial growth of INGOs in China stalled during the turbulent years of the 1930s, and then dropped precipitously with the founding of the People's Republic of China in 1949. This drop was due in part to the Interim Provision on the Registration of Social Organizations, adopted in October 1950, which reiterated the position of the Constitution in terms of citizens' rights, but also emphasized that all counterrevolutionaries would be punished. This law established the "mass organizations," such as the Women's Federation and Youth League, while effectively disallowing both domestic and international NGOs, including the medical missions. Although some INGOs left

[7] For a brief history of the Pasteur Institute in China, see http://www.pasteur.fr/recherche/unites/REG/archives_HKUPRC/History_HKUPRCL.html.

[8] Details on the early history of the Rockefeller Foundation in China can be found at: http://archive.rockefeller.edu/collections/rockorgs/cmbny.php.

[9] Robert P. Weller, *Discovering Nature: Globalization and environmental culture in China and Taiwan* (Cambridge: Cambridge University Press, 2006), 20.

China and ended their China programs, others, such as the Ford Foundation, promoted engagement with China from afar. Specifically, the Ford Foundation provided funding for China-related research centers at top universities such as Harvard, Berkeley, and Michigan, as well as academic associations such as the Social Science Research Council and Association of Asian Studies.[10]

III. *The Rapid Growth of International Environmentalism and NGOs*

While the influence of INGOs and Western science declined rapidly in China after 1949, they continued to grow in importance in Europe and North America, especially in terms of environmental issues. Indeed, widespread industrialization intensified existing environmental problems while creating new ones, such as those arising from pesticide use. Accelerating scientific progress enabled scientists to identify both the causes of these new problems as well as their intensifying effects. Universities and INGOs such as the scientific institutes mentioned previously contributed to this growing understanding of the environment, and the number of NGOs dedicated primarily to preserving the environment grew rapidly in response to emerging environmental problems.

Though NGOs concerned with the environment had been around since the founding of the aforementioned Sierra Club and various birdwatching societies, their number and scope of activities grew rapidly after World War II. These NGOs were dominated by naturalists and natural scientists, who advocated for the preservation of wildlife and natural habitats through the creation of nature reserves such as those in the United States national park system. The Nature Conservancy, founded in 1951, represented a private counterpart to the national and state park systems. The Nature Conservancy utilized the rapidly developing science of ecology to identify important natural and wildlife habitat sites, and then bought the land in order to preserve it. Tamiotti and Finger hold that this first wave of environmental NGOs was characterized by conservation ecology.[11] In other words, these NGOs held

[10] *Ford Foundation and China* published in 1991 and available online at: http://www.fordfound.org/elibrary/documents/0276/toc.cfm.

[11] Ludivine Tamiotti and Matthias Finger, "Environmental Organizations: Changing Roles and Functions in Global Politics," *Global Environmental Politics* 1, no. 1 (2001): 56–76.

to a vision of nature separate from society, and one in which scientists should be allowed to conduct research in order to determine areas most in need of preservation.

At the international level, the formation, in 1948, of the International Union for the Protection of Nature, is perhaps most representative of not only the dominant approach to environmental protection but also the role of NGOs. Renamed the International Union for the Conservation of Nature (IUCN) in 1956, IUCN is unique in that its members include both governments and NGOs. IUCN did not advocate for the environment as much as it gave scientific advice to governments. The World Wide Fund for Nature (WWF) was founded in 1961 in order to raise funds for and carry out conservation projects beyond the scope of IUCN's limited mandate. During this early period IUCN, WWF, and other environmental NGOs operated at the international level primarily by giving government officials scientific advice. They were able to influence certain international negotiations because their advice was seen as objective in that it was based on scientific evidence rather than the strategic concerns of a particular nation-state. Their scientific expertise and international influence began to grow as NGOs, such as the Rockefeller and Ford Foundations, began to contribute money to environmental concerns.

IV. *The Emergence of Global International Environmental Governance and China's Environmental Politics*

The close relationship between ecology as science and its roles in environmental protection was initially promoted within the United Nations by the United Nations Educational, Scientific, and Cultural Organization (UNESCO).[12] UNESCO arguably started what we now call global environmental governance when it organized what is now known as the "Biosphere Conference" in Paris in 1968, the first intergovernmental conference that attempted to reconcile conservation with natural resource use.[13] Over three-hundred delegates from sixty countries and international organizations attended the conference, the vast

[12] Tamiotti and Finger, 58.

[13] Official title: The Intergovernmental Conference of Experts on the Scientific Basis for Rational Use and Conservation of the Resources of the Biosphere. http://www.unesco.org/mab/mabProg.shtml.

majority of them scientists. The Biosphere Conference not only led to the formation of UNESCO's Man and Biosphere Reserve system in 1970, delegates at the conference highlighted the need to hold another, more inclusive conference in 1972.

The resulting United Nations Conference on the Human Environment (UNCHE), otherwise known as the "Stockholm Conference," was a watershed event not only for global environmental politics in general, but for Chinese environmental politics in particular. Although the Stockholm Conference took place only four years after the Biosphere Conference, its political and economic contexts were dramatically different, as were the conference's proceedings. First, the Stockholm Conference was much larger, with delegates from 113 countries and around five hundred INGOs in attendance. The level of INGO participation and influence was unprecedented, as were the kinds of INGOs involved. A number of new environmental NGOs had formed in the years separating the two conferences, many of whom were critical of industrialization, governments, and corporations. Groups began to coalesce over the issues of nuclear weapons and power, with Friends of the Earth (1969), Greenpeace (1971), and the Natural Resources Defense Council (1971) all forming during this time. Ideologically, these organizations marked the beginning of "political ecology," which held that scientific approaches are inadequate because environmental problems are rooted in political, economic, social, and cultural systems.[14] This new wave of organizations formed what we now refer to as the "environmental movement."

The ideological differences between environmental NGOs were overshadowed, however, by the differences between the industrialized countries of Europe and North America and countries in Asia, Africa, and Latin America. Governments from industrialized countries wanted to focus on issues such as transboundary air and water pollution, and argued that it was in the best interest of the Global South[15] to protect their environment. Representatives from the Global South, however, saw industrial pollution as the responsibility of the industrialized countries.

[14] Tamiotti and Finger, p. 60.

[15] Various terms have been used to refer to former colonies in Asia, Africa, and Latin America, including "underdeveloped," "the Third World," "developing countries," and "Global South," all of which reflect their political time as well as evolving theoretical understandings. No term is adequate considering the historical scope of this paper, so I have chosen to use "Global South" for convenience and consistency, and likewise use the "West" to signify North America and Western Europe.

Furthermore, for them the largest environmental issues were adequate drinking water and food. Though the division between the West and Global South took place before and during the Conference, the ideological divisions of the Cold War kept the Soviet Union and its allies from even attending.

Unlike the Soviet Bloc, however, China had begun to re-establish political relations with the West, beginning with rapprochement with the United States in 1971 and the assumption of the Chinese seat in the United Nations in 1972. Given the number of governments in attendance, the Stockholm Conference served as the perfect stage for China to re-enter world politics. Chinese leadership was also spurred to participate in the environmental conference by a number of unprecedented environmental events, including massive fish die-offs. Premier Zhou Enlai organized a group of scientists and engineers, including Qu Geping, to attend the Conference, where they were very active participants.[16] By joining with the Group of 77 (G77) of developing countries, China was able to help the G77 achieve a number of important victories during the negotiations, including having the secretariat for the United Nations Environmental Program located in the Global South, a first for a UN organization. Also, China was able to attain the inclusion of a number of principles in the final declaration, including the need to work collectively to fight pollution and the importance of free technology transfers to the Global South.[17]

The Stockholm Conference marked a turning point for China in terms of environmental protection and cooperation. The group that went to Stockholm returned home to share what they had learned at the Conference at the first National Conference on Environmental Protection in 1973, and then helped establish the Environmental Protection Leading Group in 1974, the first of its kind at the State Council level.[18] Furthermore, UN conferences and other international meetings would become one of the most important ways for Chinese scientists and engineers to learn about international technologies and knowledge related to the environment. Indeed, while a formal participant in only one International Government Organization (IGO) in 1971, China

[16] Qu Geping and Li Jinchang, *Population and Environment in China* (Boulder: Lynne Reinner, 1994).

[17] Elizabeth C. Economy, *The River Runs Black: the Environmental Challenge to China's Future* (New York: Cornell University Press, 2004), 94.

[18] Qu Geping and Li Jinchang.

had joined twenty other IGOs by 1976.[19] China also sent a number of scientists and government officials abroad to study, including Qu Geping and Song Jian. The Ford Foundation furthered the integration of international knowledge and practices concerning the environment by funding the research of the Chinese Academy of Science and Chinese Academy of Social Science.[20]

V. *Opening Up to International Foundations and Environmental NGOs: The Ford Foundation and WWF*

As with other aspects of China's political economy, environmental governance changed dramatically with the advent of Deng's opening and reform policies. Officials such as Qu Geping used what they had learned abroad to formulate the country's first *Environmental Protection Law* and *Forest Law*, both ratified in 1979. These established a legal and administrative framework for environmental protection as well as the designation of numerous nature reserves. Internationally, China continued to engage with IGOs such as the UNDP as a way to receive scientific training and technology. For example, by joining the World Health Organization's Global Environmental Monitoring System in 1980, China was able to substantially improve its monitoring technology.[21]

China also began opening its doors to INGOs. By participating in the Stockholm Conference as well as subsequent international meetings, Chinese scientists and leaders had come in contact with an increasing number of INGOs. Seeking more expert knowledge, the Chinese Academy of Science's Institute of Zoology invited the International Crane Foundation's (ICF) co-founder George Archibald in 1979 to discuss crane conservation and ways that ICF could collaborate with Chinese ornithologists. As will be discussed below, even though ICF

[19] Jonathon Harrington, "'Panda Diplomacy': State environmentalism, international relations, and Chinese foreign policy," in Harris, Paul G. (ed.), *Confronting Environmental Change in East and Southeast Asia* (London: Earthscan/United Nations University Press, 2005), 107.

[20] Guo Yong, "Case Study of the Ford Foundation's Poverty Relief Work in China," From *China National Poverty Relief Information* (Poverty Relief Office at the State Council).

[21] Michael Molitor, *International Environmental Law: Primary Material* (Boston: Kluwer Law and Taxation Publishers, 1991), 353–363.

still does not have an office in China, they have carried out a number of projects, including what has become a model conservation and development project at Caohai Lake, in Guizhou Province.

Also in 1979, WWF signed an agreement with the Chinese government to conserve the giant panda, a species that not only had become a national symbol but whose numbers were declining rapidly. WWF then sent George Schaller, already a world-famous conservation biologist, to China to study the giant panda. Schaller spent most of the next five years studying the giant panda's ecology and behavior with Chinese scientists, first in the Wolong Nature Reserve and then in the Tangjiahe Nature Reserve, both in Sichuan Province.[22] WWF also provided the funding and technical expertise necessary to conduct surveys, develop habitat protection measures, and establish a giant panda breeding center in the Wolong Nature Reserve.

During this time, WWF and INGOs were not allowed to establish offices on the mainland, so WWF set up an office in Hong Kong in 1981 in order to raise money to preserve the giant panda and other endangered species in Hong Kong and on the mainland.[23] In 1984, WWF Hong Kong began managing the Mai Po wetland reserve, transforming it into an environmental education and training center. WWF used the Mai Po Education Center, along with field studies, to provide training to China's newly appointed nature reserve managers and other government officials concerned with forestry and environmental issues. After Schaller's study of the giant panda finished, he went on to study the endangered species and habitats of the Tibetan plateau, including the Tibetan Antelope. His former Chinese and WWF colleagues were then joined by a new generation of conservation biologists, many of whom had received WWF training. They expanded their studies of the giant panda to include other regions, and began working on breeding the giant panda in captivity and formulating a national conservation strategy for the endangered species and its habitat.

The Chinese government signaled its willingness to work with more INGOs in 1985 when it established the Office of International NGO

---

[22] George B. Schaller, *The Last Panda* (Chicago: University of Chicago Press, 1994).

[23] The history of WWF in China and Hong Kong is taken largely from their homepages: http://www.wwfchina.org/english/loca.php?loca=60 and http://www.wwf.org.hk/eng/aboutus/.

Liaisons. This office sought to both encourage and manage exchanges with INGOs as a way to gain both financial support and expertise so as to carry out a number of programs related to environmental protection. Despite the formation of this office and a gradual increase in scientific exchange programs, however, the presence of INGOs in China was very limited throughout the 1980s.

China's openness to INGOs reached a new level when the Ford Foundation was allowed to set up an office in Beijing in 1988, under the auspices of its sponsor, the Chinese Academy of Social Science. The Ford Foundation enjoyed unprecedented access to government institutions, in part because many of the scientists and officials who had benefited from the Ford Foundation's fellowship programs had returned to take leadership positions in the government. In the beginning, the Ford Foundation focused on the environment and development, legal system reform, rural poverty and resources, international affairs, and urban poverty.[24] The Ford Foundation had invested nineteen million USD on environmental conservation, poverty relief, and other programs up to 1988, and expanded the scope of its work after establishing the Beijing office. From 1988 to 2005, the Ford Foundation gave another 207 million USD in grants for a variety of cross-cutting programs, including NGO capacity-building and the training of nature reserve staff and other Forestry Bureau officials.[25]

The impact of the Ford Foundation in China, however, cannot be measured by the amount of grant money distributed. First and foremost, the Ford Foundation, along with WWF, has played a crucial role in importing international ideas about development and environmental conservation practice. For example, the Ford Foundation trained Yunnan forestry officials in participatory resource management practices.[26] This and similar training conducted by WWF enabled forestry officials to successfully apply for other international conservation funds because they could frame their projects in the language used by international development institutions.

[24] *Ford Foundation and China*, published in 1991 and available online at: http://www.fordfound.org/elibrary/documents/0276/toc.cfm.

[25] Information on later Ford Foundation activities in China was gathered from annual reports and the Beijing office website: http://www.fordfound.org/global/office/index.cfm?office=Beijing.

[26] Kenji Kitamura and Guangxia Cao, "Community Forestry in Yunnan Province," *China Environment Series* 6 (2005): 116–119.

Second, the establishment of the Ford Foundation office in Beijing was a test-case of sorts, paving the way for other INGOs to open offices in China, including other foundations and environmental NGOs. The Ford Foundation actively encouraged other INGOs to enter China by subcontracting some of their projects to other organizations. For example, rather than managing a proposed forestry-conservation project in Yunnan, the Ford Foundation brought in Winrock International, a development NGO that has worked in China ever since.[27]

VI. *China Opens its Doors Further: Environmental Diplomacy Post-Tiananmen*

Soon after the Ford Foundation opened its Beijing office, China's leaders became even more interested in cooperating internationally on environmental issues. This was not due to anything the Ford Foundation did, but rather was a reaction to the diplomatic isolation that followed the Tiananmen incident. Western leaders repeatedly refused to meet with their Chinese counterparts in the months that followed. Just as the Stockholm Conference had done in the early 1970s, the discussions leading up to the United Nations Conference on Environment and Development, or Rio Earth Summit, in 1992 provided a safe way for Chinese and Western leaders to reconnect. Just before attending the Earth Summit, the State Council formed the China Council for International Cooperation on Environment and Development (CCICED), a very high-level non-governmental advisory body.[28] The CCICED brought together high-level officials from various ministries and international experts, such as prominent professors and the heads of INGOs such as WWF. The Chinese government was very active at the Earth Summit, becoming the first major country to ratify an Agenda 21 plan afterwards. In the years preceding and following the Rio Earth Summit, China co-hosted numerous international environmental conferences with the World Bank, UNEP, and WWF.[29]

---

[27] Winrock International's homepage for details of their projects in China: http://www.winrock.org/index.asp.

[28] Details on the history, working groups, and projects of the CCICED can be found at http://www.harbour.sfu.ca/dlam/history.html.

[29] A. I. Johnston, "China and International Environmental Institutions," in M. McElroy, C. Nielson, and P. Lydon (eds.), *Energizing China: Reconciling environmental protection and economic growth* (Newton, MA: Harvard University Committee on Environment, 1998).

Perhaps more importantly for China's environmental situation, however, was the large number of local, national, and international NGOs present at the Earth Summit. In many ways, the Earth Summit, coming right after the end of the Cold War, was a watershed event for NGOs around the world.[30] The Chinese delegation was embarrassed by the lack of Chinese NGOs, something which made it easier for Friends of Nature, Global Village of Beijing, and other environmental NGOs to form in the following years. Chinese leaders also became more welcoming of INGOs. Indeed, at the time of the Earth Summit there were less than ten INGOs working on nature conservation in China. This number grew from twelve in 1995 to almost thirty in 2000, with the number of INGOs with actual offices rising from single digits in the mid-1990s to almost twenty by 2000.[31]

VII. *INGOs, Capacity Building, and the Environmental Conservation Community*

The growth of INGOs and domestic environmental NGOs were the products of the same processes. Namely, the Chinese government allowed more INGOs to enter China and the first domestic NGOs to form. INGOs with longer histories in China in turn provided financial and institutional support so that domestic environmental NGOs could establish themselves. Moreover, the community was small enough in the early 1990s that INGOs, domestic NGOs, and related government agencies, such as local forestry bureaus, often shared close relationships. For example, individuals who received wetlands-management training from WWF or forestry-conservation training from the Ford Foundation either returned to the forestry bureau, joined an INGO, or started their own NGO. While their institutional affiliation may have changed, they had either worked with or shared training sessions with counterparts in other organizations.

A good example of the ways in which INGO staff moved between organizations is Jim Harkness. Harkness first visited the Caohai Nature Reserve in Guizhou Province as part of a research group from the

[30] Tamiotti and Finger.

[31] Wu Fengshi, "Double Mobilization: Transnational Advocacy Networks for China's Environment and Public Health" (Ph.D. diss., unpublished, University of Maryland, 2005), 146.

International Crane Foundation in the mid-1980s. After receiving his Masters from Cornell University he returned to China as a program officer for the Ford Foundation. Harkness then became the China Representative for WWF as it established its China Programme Office in 1996, a position he held until 2005. Similarly, Dr. Lu Zhi conducted intensive research on the giant panda in the 1980s before becoming a WWF program officer and Peking University biology professor. She now heads Conservation International's China office.[32]

Others, such as Dr. Xu Jianchu, graduated from INGO training programs to start their own NGOs. Dr. Xu was working at a research station in Yunnan when he was chosen to be a part of a Ford Foundation program that sent researchers to universities in the Philippines and Thailand. After returning to China, the Ford Foundation and WWF provided financial support for Dr. Xu to start the Center for Biodiversity and Indigenous Knowledge (CBIK) in Yunnan. CBIK now has around fifteen full-time staff, supports the research of foreign and Chinese scholars, and helps Chinese students get advanced degrees abroad.[33] Dr. Xu has continued to employ the participatory principles he learned through the Ford Foundation program while working to document indigenous knowledge, promote community governance of resources and economic development, and protect the biodiversity of Yunnan. The connections that Harkness, Dr. Lu, and Dr. Xu made while working with various NGOs and government agencies enabled them to carry out a number of projects that would not have been possible as recent as the early 1990s.

VIII. *International Funding Priorities and Local Ecosystems: the Case of Southwest China*

While China's central government has played a very large role in determining which INGOs have entered China, international funding priorities have also been extremely important regarding which INGOs have come and when. For example, in the late 1980s, interest in Tibetan culture spurred INGOs such as the Mountain Institute to come to China, while the late 1990s saw the influx of INGOs con-

[32] Wu Fengshi, 180.

[33] Information on CBIK taken from their website, which includes ample documentation in Chinese and English on their various projects: http://www.cbik.org/.

cerned specifically with biodiversity and climate change. No area in China has benefited more from international funding priorities, however, than Yunnan Province, in southwest China. Yunnan is located on the eastern edge of the Himalayan Mountains, serving as the northern border for Laos and Vietnam. As such, Yunnan has many of the same cultural and ecological traits as both Tibet and Southeast Asia, and thus benefited from projects focusing on these two areas. For example, the Ford Foundation launched its first Chinese forest conservation projects there because it thought it could build on its experiences in Thailand.[34] Later, MacArthur Foundation projects focusing on the eastern Himalayas provided funding for research institutes and universities in Yunnan along with Tibet, Qinghai, and Sichuan Provinces. Similarly, the international focus on tropical rainforests in the early 1990s meant that WWF received donations that could only be used to protect rainforests. Yunnan again benefited as it is the only location in China other than Hainan Island that has rainforests.

While international funding priorities played a part, the Yunnan Provincial government was also assertive in attracting INGOs. The provincial government was correct in thinking that once a few INGOs set up offices in the provincial capital of Kunming, others would follow. The provincial government's openness to INGOs in the early 1990s was in stark contrast to Beijing, where it was much more difficult for INGOs to open offices. Indeed, many INGOs did not establish any offices in China until the mid- to late-1990s because of the difficulties in registering. Even WWF, perhaps the most active INGO in China in the 1980s, did not establish its China Programme Office until 1996.

The initial openness to INGOs and international ideas enabled Yunnan to become a hub for a wide range of INGOs working in Southwest China, including those working on poverty alleviation, education, women's rights, and HIV/AIDS. The institutional infrastructure built in the early 1990s, in addition to it being China's most biologically diverse province, enabled Yunnan to receive millions of dollars earmarked for environmental protection. For example, when The Nature Conservancy (TNC) came to China in 1998, it signed an agreement to invest two million USD and provide technical assistance in establishing the Yunnan Great Rivers Project.

---

[34] Discussion of Yunnan is taken largely from: Jim Harkness and Nick Young, "What about Guizhou? Reflections on cooperation prospects," *China Development Brief* (2005) http://www.chinadevelopmentbrief.com/node/99.

Yunnan and other provinces in southwest China received even more international attention, however, after the publication of *Hotspots: Earth's Biologically Richest and Most Endangered Terrestrial Ecoregions* in 1999. The mountains of southwest China were identified as one of the world's twenty-five biodiversity hotspots. Conservation International (CI) thus came to China in 2000, and in 2001 began administering the funds for a consortium of international agencies and governments that included the World Bank, the MacArthur Foundation, and the government of Japan called the Critical Ecosystem Partnership Fund (CEPF).[35] CEPF has pledged six million USD to protection of the biodiversity of the southwest region, while TNC now has an annual China budget of around five million USD. WWF, TNC, and CI carry out a variety of small- and large-scale projects aimed at building the capacity of local groups to monitor, research, and protect the biodiversity of their areas.

These areas are home to a large number of ethnic minorities, the cultural diversity in many ways matching the biological diversity. These areas are also some of the poorest in China. Therefore, many of the projects combine environmental protection with economic development and overall capacity building. Some of these projects are modeled on an integrated conservation and development project conducted at the Caohai Nature Reserve in nearby Guizhou Province. This project has been a joint collaboration of the International Crane Foundation, Trickle Up!, a micro-finance INGO, and the local nature reserve staff. Given its importance as a model for nature conservation in China, I will discuss it in some detail below.

IX. *The International Crane Foundation: From Scientific Collaboration to Broad-Based Cooperation*

Though WWF has been the most active nature conservation INGO in China over the past twenty-six years, and TNC and CI now have very large China budgets, there have been a number of other INGOs concerned with biodiversity. Some of these INGOs, such as Save China's Tigers, International Snow Leopard Trust, and the Wild Camel Protection Association focus on one species. Others focus on a single

---

[35] Information on CEPF is largely derived from its website: http://www.cepf.net/xp/cepf/about_cepf/index.xml.

kind of habitat, such as the International Rivers Network, The Mountain Institute, and Wetlands International. They have all contributed to the protection of biodiversity in China, but the International Crane Foundation deserves special mention because of its long history and innovative projects.

Although the ICF is a much more specialized environmental INGO than WWF or TNC, its experience is in some ways paradigmatic of the evolution of INGO work in China over the last three decades. As mentioned previously, ICF co-founder George Archibald was invited by the Chinese Academy of Science's Institute of Zoology in 1979 to discuss the status and protection of cranes. Since that time, ICF has not only worked in the Caohai Nature Reserve mentioned above, but has also the Northeast, the Yangtze River basin, and other crane breeding and wintering sites. Not only has ICF worked in various locations around China, they have had numerous government, academic, and NGO partners. I believe that the number and variety of their partners is indicative of the multi-layered partnerships that environmental INGOs are forming in twenty-first-century century China. Whereas Dr. Schaller, from WWF, was confined to a few locations and government partners when he studied giant pandas in the early 1980s, WWF, TNC, and others work with various partners on various levels throughout China.

Table 10.1 details some of ICF's partners over the last six years, and Table 10.2 details their recent and ongoing projects.[36] As these tables show, ICF has numerous ongoing projects, many of which are a combination of environmental education, ecotourism, and research on cranes and their habitats. The publication of scientific and educational materials has already done much to raise awareness about protecting cranes and the wetlands they depend on. Indeed, I have met numerous graduate students, researchers, and nature reserve staff that regularly use *China Crane News*. The ICF is thus supporting a generation of scientists not unlike those who worked on the WWF's efforts in regard to the giant panda in the 1980s. Similarly, the educational materials that ICF has made in conjunction with the Chinese NGO the Beijing Brooks Educational Center are on the bookshelves of numerous NGOs and environmentally-minded teachers throughout China.

---

[36] Information listed in the tables was compiled from ICF's website and various documents on the Woodrow Wilson Center's China Environment Forum website: http://www.wilsoncenter.org/index.cfm?topic_id=1421&fuseaction=topics.home.

X. *The "Caohai" Model: From Conservation Conflict to Integrated Conservation and Development*

Though ICF's educational materials have been distributed throughout the flyways of China's crane species, ICF arguably achieved its greatest impact through its innovative Integrated Conservation and Development Project (ICDP) at the Caohai Nature Reserve. After WWF staff and government officials identified Caohai Lake in Guizhou Province as the wintering site for the endangered Black-necked Crane, the lake and surrounding area were declared a nature reserve in 1985. In order to ensure the survival of the cranes, activities that threatened the cranes, such as hunting and fishing, were forbidden within the nature reserve.[37] The problem with this approach, however, was that twenty-five thousand people lived around the lake and depended on it for their survival. Indeed, the communities around Caohai Lake were some of the poorest in all of China. Not surprisingly, conflicts between local residents and nature reserve staff escalated rapidly. In order to diffuse these conflicts and protect the Black-necked Crane, ICF decided to intervene. ICF joined with Trickle Up!, a New York-based microfinance program, and the nature reserve staff to design a program that would enable the local communities to develop economically while preserving the cranes.

ICF and Trickle Up!, with funding from the Ford Foundation, launched their ICDP in 1993. They established a microfinance system whereby groups of three to five villagers would receive a 100 USD grant in order to start a business, on the condition that it did not hurt the cranes. They combined the microfinance program with environmental education campaigns about soil erosion and participatory planning techniques, and provided information on the cranes and their protection. After two years, they expanded their program by establishing Community Trust Funds (CTF) which included ten to fifteen families. The CTFs used participatory planning methods when deciding how to manage and use their funds. Since establishing CTFs, the villagers' incomes have increased and their attitudes towards conservation have

[37] The discussion of the Caohai Lake Nature Reserve is based on the International Crane Foundation website, www.savingcranes.org, and on Melinda Herrold-Menzies, "Cranes and Conflicts: NGO Programs to Improve People-Park Relations in China and Russia" (Ph.D. diss., unpublished, ESPM, University of California, Berkeley, 2002).

changed. There is still considerable human pressure on the lake, and thus the project has not been a complete success from the viewpoint of conservation. However, the ICDP has transformed what was previously a very contentious relationship between the villagers and the reserve staff, and between the villagers and the cranes.[38] The "Caohai Model" is now being replicated in various nature reserves around China.

Table 10.1: International Crane Foundation project partners

National Government	State Forestry Administration of China
Provincial Forest Administrations	Heilongjiang, Jilin, Yunnan, Jiangxi, and Guizhou Provinces
State-controlled Research Centers	Chinese Academy of Sciences, Kunming; National Bird Banding Center, Crane and Waterbird Specialist Group of the China Ornithological Society
Nature Reserves	Poyang Lake, Zhalong, Naoli River, Cao Hai, Xianghai
Domestic NGO and Academic Partners	Beijing Brooks Education Center, Yunnan Normal University, Tibet Plateau Institute of Biology
International Governmental Organizations and INGOs	Convention on Conservation of Migratory Species of Animals, Cracid Breeding and Conservation Center (Belgium), Global Environment Facility (GEF)/UN Environment Programme (UNEP), WWF-China
Foundations and other Funding Sources	The Henry Luce Foundation, Trickle Up Program, Liz Claiborne and Art Ortenberg Foundation, Ford Foundation, John D. and Catherine T. MacArthur Foundation, Keidanren Nature Conservation Fund, Hamill Family Foundation, Brehm Fund for International Bird Conservation, Felburn Foundation, U.S. Department of State, ICF member donations, individuals.

[38] Melinda Herrold-Menzies, "Integrating Conservation and Development: What We Can Learn From Caohai, China," *Journal of Environment and Development* 15, no. 4 (2006).

Table 10.2: International Crane Foundation's current projects in China

Project Name	Stated Purpose	Timeline
Conservation of Globally Significant Wetlands Used by Siberian Cranes	Biodiversity, Conservation Management	Initiated March 2003; Targeted Completion 2009
An Environmental Education Process that Involves Local Communities with Wetland Reserve Management in China	Environmental Education	Initiated 2003; Targeted Completion 2006
Integrating Conservation with Rural Development at Caohai Nature Reserve	Biodiversity, Conservation Management	Initiated 1993; Ongoing
Migration Study of Black-Necked Cranes in Southwest China	Migration Study, Wintering and Breeding Ecology, Environmental Education	Initiated 2004; Ongoing
Promoting Ecotourism in Tibet Through a Culturally Responsive Conservation Education Program and Professional Training	Ecotourism, Environmental Education, Economic Development	Initiated 2003; Ongoing
Publication of China Crane News	Biodiversity, Environmental Education	Initiated 1997; Ongoing
Studies of Waterbirds, Water Levels, and Aquatic Food Plants of Threatened Wetlands at Poyang Lake	Biodiversity, Conservation Management	Initiated 1999; Ongoing

XI. *Beyond Biodiversity and Environmental Education: Civil Society*

From the 1980s through the mid-1990s, the vast majority environmental INGOs in China worked to protect biodiversity, promote environmental awareness, and build the capacity of Chinese scientists, nature reserve staff, and later NGOs. China's extraordinarily rapid economic growth, however, has not only led to a number of new environmental problems domestically, but has also raised the nation's environmental profile internationally. The last decade has thus seen an influx of INGOs working on other environmental issues, including the empowerment of Chinese civil society. While many of the earlier INGOs conducted capacity-building campaigns, most were not explicitly focused on civil society. The growth of programs focused on civil society is due to a combination of factors.

First, following the Rio Earth Summit, the Chinese government became more accepting of NGOs and the concept of civil society, with some leaders openly discussing the role of civil society in a "harmonious society." Furthermore, nearly three decades of rapid economic growth has produced a growing middle class. Although a growing middle class has been associated with environmentally-friendly values in many countries in the world, it is always associated with higher levels of consumption. Thus, small INGOs and foundations such as Global Greengrants Fund (GGF) and ECOLOGIA have been working to support grassroots environmental NGOs and student groups throughout the country, by providing small grants for activities and publications.[39] GGF and ECOLOGIA, while providing far less money to domestic NGOs than WWF and CI, have nonetheless played a vital role by spreading the money among groups that would have floundered otherwise. While GGF and ECOLOGIA continue to support student groups and grassroots NGOs, the American Bar Association (ABA) is working with the State Environmental Protection Agency (SEPA) and training Chinese lawyers through its environmental governance program.[40] Through seminars and workshops, the ABA hopes to raise awareness about environmental rights among both lawyers and citizens alike.

XII. *Sandstorms and East Asian Cooperation*

Unconnected to the efforts of American and European INGOs, Japanese and Korean NGOs have been coming to China in increasing numbers over the last fifteen years. Possibly because of language barriers and different priorities, Japanese and Korean NGOs do not work closely with other INGOs, nor do they work in the INGO "hotspot" of southwest China. Rather, these NGOs work in the northern provinces of Inner Mongolia, Shanxi, Ningxia, and Gansu. They are primarily concerned with desertification and the resulting sandstorms that began drawing significant media attention in the late 1990s.

[39] See the Global Green Grant Fund for details of their grantmaking: http://www.greengrants.org.

[40] See the ABA website for details: http://www.abanet.org/aba-asia.

Japan's Green Earth Network (GEN) has been working around Datong, Shanxi Province, since 1992 when its founder, Kunio Takami, began spending over three months a year on the Loess Plateau. Recognizing that for an environmentally-motivated project to be successful it would have to work with the local government and address the needs of the local residents, GEN focused on methods that would generate income *and* protect the fragile soils around Datong.[41] After experimentation, Takami found that apricot trees are resistant to drought and that farmers growing apricots could earn ten thousand yuan (one thousand three hundred USD) annually, compared to two thousand yuan (two hundred and sixty USD) for other crops. Over the last fifteen years, GEN has raised almost three million USD and recruited almost twenty-five hundred Japanese volunteers to help tens of thousands of local people to plant over seventeen million apricot and pine trees. Another Japanese NGO, the Organization for Industrial, Spiritual and Cultural Advancement International (OISCA) has been conducting afforestation programs along the Yangtze River and in Inner Mongolia since 2000. OISCA established a center for technical cooperation with Shanghai's Jiaotong University in 2002 and founded the Alxa Desert Ecological Research and Training Center at its Inner Mongolia site in 2006.[42]

South Korea's Northeast Asian Forest Forum (NEAFF) was established in 1998 by former forestry officials and scientists, and held its first international seminar on desertification in 1999. NEAFF has worked with forestry bureau officials on reforestation projects in Inner Mongolia and Shandong provinces. NEAFF has also worked with Korea's Forestry Service and China's Forestry Administration to hold two to three conferences a year for scientists and government officials since 1999. The Korean Federation for Environmental Movement (KFEM), the largest and most prominent environmental NGO in South Korea, has been working with China's Friends of Nature and Echoing Steppe since 2003. They have organized a number of educational exchange programs and fieldtrips for Korean and Chinese high school and university students. KFEM also has hosted annual meetings in China and Korea to help NGOs network domestically and internationally on desertification and other regional environmental issues. Feeling that the emphasis on trees

[41] Shi Pengyun, "Saving someone else's backyard," *China Daily*, April 13, 2007.

[42] Robert Efird, "Japanese NGOs in China," *China Development Brief* (January 2007), 11–15.

has been misplaced, KFEM and their Chinese partners have been co-managing hybrid grass test sites with local herders and farmers in four locations in Inner Mongolia and Jilin provinces. Finally, KFEM has helped Korean reporters and documentary filmmakers come to China in order to raise awareness of the economic and cultural issues that lead to desertification.

XIII. *Climate Change, Energy Conservation, and Alternative Energy*

Although the vast majority of INGOs in China have been mainly concerned with biodiversity protection, an increasing number of INGOs are coming to China to promote energy conservation, alternative energy such as solar power, and cleaner ways to produce energy. The emergence of climate change as possibly the most serious global environmental issue, and China's rapidly-increasing greenhouse gas emissions, have made China's rapid industrialization and corresponding energy use very important topics among many INGOs in Europe and America. Indeed, some feel that helping China curb its greenhouse gas emissions is the single greatest challenge facing the world today. For example, the EcoLinx Foundation was founded in 2002 with the sole purpose of helping Chinese leaders develop less fossil-fuel-intensive development strategies.[43] This INGO is very small and focuses mainly on connecting with Chinese academics to strategize about energy policy.

Though EcoLinx is too small to carry out large-scale projects, the Energy Foundation (EF) has enough funding to redesign the transportation systems of entire Chinese cities. EF, which is funded by the David and Lucile Packard Foundation and The William and Flora Hewlett Foundation, has been working since 1999 on a range of programs aimed at increasing energy efficiency and reducing overall greenhouse gas emissions. EF conducts a variety of policy-relevant studies in order to better advise both local and national governments. Its sole focus on energy, along with the financial backing mentioned above, has enabled EF to address energy issues from a variety of angles.[44]

---

[43] See their website for details: http://www.ecolinx.org.

[44] Information on the Energy Foundation was gathered mainly from their website: www.ef.org/programs.cfm.

One EF project focuses on monitoring and improving the energy efficiency of small, everyday electrical appliances that will almost certainly be purchased by hundreds of millions of Chinese as their income levels rise. Other projects center on energy efficient and energy self-sufficient buildings. They are also promoting the adoption of policies and building codes that require greater energy efficiency in both appliances and buildings. Furthermore, they invest in research on how renewable energies can become more economically feasible. Their biggest project in terms of infrastructure, however, is the construction of the extraordinarily innovative Bus Rapid Transit (BRT) in Beijing, Kunming, and Xian. Realizing that the growing number of cars in China is as large an issue as power production, the Energy Foundation has invested hundreds of millions of dollars into the construction of BRT systems in these cities, with the hope that other cities will adopt the technology.

XIV. *Greenpeace, Climate Change, and Environmental Health*

China's growing energy consumption also attracted the attention of possibly the most famous INGO in the world, Greenpeace. Greenpeace, known for its anti-corporate stance and attention-grabbing activities, established an office in Hong Kong in 1997 and has since opened an office in Beijing.[45] Following Greenpeace International's lead, Greenpeace China has also made stopping global warming its number-one issue. Greenpeace China thus conducts education and awareness campaigns, as well as media-catching actions, to highlight the dangers of global warming. Greenpeace supports the phasing out of fossil fuels and the development of renewable energy sources, and continues to completely disavow nuclear power as an option.

In a return to the environmental health focus of the medical missionaries, Greenpeace China is one of the only INGOs to focus on toxic chemicals. In 2006, Greenpeace China conducted a very interesting campaign on the dangers of toxic chemicals in common electronic goods such as cell phones and computers. This campaign is especially appropriate in southeast China, one of the electronic waste recycling

---

[45] The following information was gathered through interviews and the Greenpeace China website: www.greenpeace.org/china/en.

and disposal hubs in the world. Greenpeace has been highlighting the high levels of heavy metals and other toxics in the water and land around these disposal sites.

Greenpeace has also shown how the agriculture produced on farms around electronic-waste disposal sites also contains high levels of toxins. This campaign is also connected to Greenpeace's larger mission of promoting food safety and sustainable agriculture. One of the cornerstones of this campaign is the elimination of genetically engineered agriculture products.[46] As part of this campaign, Greenpeace is raising consumer awareness about the dangers of pesticides and promoting labeling for genetically engineered foods. Indeed, some believe that Greenpeace's somewhat rocky relationship with the Chinese government improved considerably when Greenpeace loudly accused Monsanto of bio-piracy in regard to taking rice and other plants from China in order to map their genomes.

XV. *Conclusions*

International NGOs have been working on environment-related issues in China for over a century. After an early emphasis on environmental science and health, INGOs were largely absent from China during the middle of the twentieth century. As China began to experience and face environmental problems in the 1970s, however, INGOs began working with Chinese officials and scientists within the limited confines of international conferences. After Deng Xiaoping began the opening and reform period in 1978, however, INGOs began working in earnest with Chinese scientists on the mainland. Although these early collaborations focused largely on protecting a few key species such as the giant panda, Chinese officials and INGOs began to earn one another's trust. The establishment of forestry and wetland conservation training programs in the 1980s educated a generation of scientists, officials, and nature reserve staff about internationally-used methods. These individuals then began to either work for INGOs or start their own NGOs.

After growing slowly through the 1980s and early 1990s, the number of INGOs and domestic NGOs working on environmental issues

[46] These products are often lumped in with other Genetically Modified Organisms, or GMOs.

in China has grown rapidly. INGOs not only support and work with NGOs but also continue to work with state officials at the national, provincial, and local levels. Together, INGOs, domestic NGOs, and Chinese government agencies are working to build an environmentally harmonious society. China's unprecedented economic expansion, however, is creating new environmental problems while exacerbating the old problems of habitat and biodiversity loss. INGOs and domestic NGOs are thus expanding the scope of their activities beyond nature conservation and environmental awareness to include energy-related issues and environmental health issues such as desertification and concomitant sandstorms, water pollution, and toxic waste. Finally, an increasingly wealthy and educated middle-class is emerging; their environmental awareness provides hope for China's future.

PART TWO

ECOLOGY

This part of the yearbook discusses major events that have occurred or are likely to occur in China's ecological development.

The article entitled "Are Fences and Grazing Bans the Best Tools for Controlling Desertification?" raises questions about an enormous project. According to the Chinese Academy of Social Sciences, twenty years of this policy has yielded less than satisfactory results. Well-managed herding can actually help grasslands; bovines, camels, and other large, grass-eating animals have long been part of the grassland ecology, and the enclosure system disrupts this connection. Liu Shurun, the article's author, proposes an alternative policy: returning land ownership to native herdsmen. By moving to a private ownership model, much like the one used to manage rangeland in Western countries, owners accrue the benefits of proper land management. They also bear the risks of improper resource management. Perhaps we should learn from, and apply, this model to improve grassland use and maintenance.

The western section of the South-to-North Water Diversion Project will also impact grasslands and surrounding areas. The proposed route cuts through mountainous terrain, rising as high as three thousand five hundred meters above sea level as it passes through the eastern Qinghai-Tibetan Plateau. This route will access the upper and middle reaches of six major rivers: the Yarlong Tsangpo, Nujiang,[1] Lancang,[2] Jinsha, Yalong, and Dadu.[3] Because mountainous ecological systems are especially fragile, the international community deems this construction project as risky, at best. Unsurprisingly, a project of such scale and potential influence has drawn widespread attention within China and across the world.

---

[1] Translator's note: The Nujiang is also known as the Salween River in India.

[2] Translator's note: Most Westerners know the Lancang River as the Mekong, which is a name used in its lower, southeasterly reaches.

[3] Translator's note: Dadu is sometimes spelled Tatu.

CHAPTER ELEVEN

THE ENVIRONMENTAL IMPACTS OF LARGE-SCALE CONSTRUCTION PROJECTS

Fan Xiao

Large-scale projects under consideration, and their possible environmental impacts, have stirred debate in Chinese society during 2006. The most heated exchanges focused on the proposed Western Line Project of the South-to-North Water Diversion Project (S2N) and the Shuotian Grand Canal Project. Furthermore, new concerns regarding ongoing projects, such as the Three Gorges Dam, have come to light, further sharpening discussions.

Key words: South-to-North Water Diversion Project, Western Line Project, Shuotian Grand Canal Project, Three Gorges Dam, environmental impact

I. *Disputes over Several Large-scale Projects in 2006*

A. *The Western Line Project of the South-North Water Diversion Project*

The construction of the Eastern and Central Lines of the South-to-North Water Diversion Project (S2N) has already encountered a series of problems. The Eastern Line suffers from severe water pollution, which requires expensive treatment; the Central Line lacks sufficient water volume. Specifically, the Danjiangkou Reservoir can only supply 9.7 billion cubic meters (m^3) of water, far from the planned 14 billion m^3. Additionally, both sections traverse seismically-active zones, which raises concerns about their long-term durability.

Beyond these technical difficulties, both sections have come up against mounting social pressure as the impact of population displacement causes more acute concerns. Many ethnic minorities will be among those most affected. Additionally, cultural relics and historic sites along both lines have been heavily damaged.

The Western Line faces especially thorny difficulties, even though its construction has not yet started. Local conditions pose serious engineering and logistical challenges: thin air, extreme cold, and lack of local infrastructure among them. Making matters worse, the water supply is sometimes eighty to five hundred meters lower than the water delivery area, which means that pumping will be necessary. The project's path crosses through vulnerable ecosystems. Mitigating the environmental impact will obviously be a challenge; convincing critics that this can be done well, or at all, only complicates matters.

The Western Line Project (WLP) calls for the construction of a series of dams, mostly between one hundred and fifty and three hundred meters high, over the primary courses of the Tongtian, Jinsha, and Dadu[1] Rivers and their tributaries. The resulting reservoirs will provide the water-storage volume necessary to charge and maintain the flow of the Diversion Project. A series of deeply-buried tunnels will serve as conduits to carry water to and between rivers, reservoirs, and canals. The longest of these will be 164 kilometers.[2]

Planning for the western part of the project began in the 1980s. After completing a thoroughgoing vetting, the Ministry of Water Resources adopted the plan in May 2001. Note, however, that the plans have not been made available to the public, experts outside the ministry, or local authorities involved in water management, construction, or other relevant forms of regulation and oversight. However, a public pronouncement in 2004 that set the construction start date for 2010 finally put the project on the public radar. And with that, public comment and debate began in earnest.

Soon thereafter, Sichuan-based experts in geology, engineering, hydroelectric engineering, ecology, economics, sociology, and anthropology weighed in on the projected impacts of the project. They distilled their findings into a May 2006 report entitled *The Memorandum of the Western Line Project of the South-to-North Water Diversion Project* (or *Memorandum*). This report scrutinized the project's geological suppositions and related planning, its likely ecological impact, and its economic feasibility, as well as its social and cultural impacts on the Tibetan region.

As the first publication representing voices critical of the S2N, the *Memorandum* has fomented comment and debate. In its wake, numerous influential newspapers and websites have presented reports and commentary on the S2N, including the *21st Century Business Herald* and

[1] Translator's note: The Dadu is also spelled Tatu.

[2] The Yellow River Planning and Research Institute under the Ministry of Water Resources, *The South-to-North Water Diversion Project Compendium and First-Phase Project Plan Brief Report, 2001.*

Eastday.com. One article posted online about the water diversion system attracted over ten thousand comments.

Commentary took more direct forms as well. Lu Jiaguo, a researcher at the Sichuan Province Academy of Social Sciences, wrote to Premier Wen Jiabao twice regarding his doubts over the feasibility, efficiency, and economic rationality of the WLP. He suggested suspending work on the project until concerns could be meaningfully addressed.

These actions had the desired impact. The Ministry of Water Resources postponed the bidding and contracting process, originally slated to begin in 2004, until further research could be completed. This has opened up discussion of alternatives to the project, such as transferring water from small rivers in the upper reaches of the Yangtze River above the Three Gorges Dam.[3]

B. *The Great Western Line Project (Shuotian Grand Canal Project Plan)*

Since the idea of a south-to-north water diversion system first emerged, more than a hundred plans have been proposed. Apart from the Eastern, Central and Western Line Projects, much recent discussion has focused on the Great Western Line Project plans. Unlike the aforementioned three lines, the Great Western Line plans divert water from multiple sources: the Yangtze River, and the Lancang,[4] Nujiang[5] and Yarlung Tsangpo[6] Rivers, which all lie west of the Yangtze. Owing to technical complications, the Great Western Line Project only exists on paper at this point.[7]

---

[3] Yang Chuanmin, "The South-North Water Diversion Project Postponed because of Complaints from the Local People," *Nanfangdushi Daily*, Dec. 5, 2006.

[4] Translator's note: The Lancang River is also known as the Mekong in Southeast Asia. The Burmese called it the Mae Khaung, the Thai refer to is as Mae Nam Khong, the Lao use the name Mènam Khong, the Khmer say Mékôngk or Tonle Thom, and the Vietnamese say Sông Cửu Long.

[5] Translator's note: The Nujiang River is also known as the Salween River in India, the Thanlwin in Thailand, and the Salawin in Myanmar. It empties into the Indian Ocean.

[6] Translator's note: This is the Tibetan name for the river. Mandarin speakers generally refer to it as the Yaluzangbu River. The river is better known as the Brahmaputra as it makes its way through India, before emptying into the Indian Ocean.

[7] Translator's note: The descriptor "Great Western Line" originated from the Shuotian Grand Canal proposed by Guo Kai and others. In fact, the idea and related study of the Great Western Line project dates back to the 1950s.

The idea of diverting water from multiple rivers dates back to the 1950s and has manifested itself in many forms:

1. The Yellow River Water Conservancy Committee's plan to divert water from the Nujiang River and Lancang River;
2. The Yangtze River Water Conservancy Committee's plan to divert water from the Nujiang River and Dadu River;
3. Hydrology engineering professor Huang Wanli's plan to transfer water from the beginning of the big bend in the Yarlung Tsangpo River directly into the Yellow River;
4. Chen Chuanyou's[8] plan to build super-sized hydroelectric power plants at the bend in the Yarlung Tsangpo River and the Yellow River, and to use part of this energy to run powerful electric water pumps to divert water toward a transfer system;
5. Guo Kai's[9] Shuotian Grand Canal plan to divert water from Shuozhou, Shanxi Province, through the Sanggan River system to Tianjin, which evolved into a slightly different project designed to divert the Yarlung Tsangpo's waters to the Yellow River.

Throughout most of this process, the various Great Western Line plans had been discussed behind closed doors by a handful of experts and government officials. However, the publication of Ling Li's *Tibet's Water Will Save China*, first published as a long investigative report in November 2005, and then in book form in April 2006, changed all that. The report and book, which served as a polemic supporting Guo Kai's audacious proposal, became an immediate hit, and its wide circulation has triggered an enormous outpouring of public opinion and debate over the advantages and disadvantages of the WLP.

Guo Kai's design is in fact an extension of the South-to-North Water Diversion Project's subsidiary Western Line. Both plans call for diverting water from the Tongtian, Yalong, and Dadu Rivers into the Yellow River. However, in Guo's proposal, dams will be built on the Yarlung Tsangpo, Nujiang, and Lancang Rivers. Watersheds between

[8] Translator's note: Chen is a research fellow at the Chinese Academy of Science's Institute of Geographic Sciences and Natural Resources Research.

[9] Translator's note: Guo is probably best known as the mastermind behind the Shuotian Grand Canal Project Plan; he is largely a self-made man who has had an interest in hydrology and water transfer systems dating back at least twenty years.

these rivers will be opened to let water flow through; these pipes will lead to tunnels and aqueducts that will then connect to reservoirs. Guo's proposal also calls for the use of controlled detonations to cause landslides that would block the rivers to be dammed. This would facilitate dam construction. Water from the resulting reservoirs will be diverted into three of the largest endorheic basins in the region, the Qaidam, Tarim, and Junngar Basins, via Qinghai Lake. From there, water will flow into the Yellow River, supplying various northern destinations such as Beijing, Tianjin, Heibei, Shanxi, Liaoning, and the grasslands of Inner Mongolia.

Heated debates have erupted over Guo's proposal. Former President Jiang Zemin, in a speech delivered on May 24, 1998, pointed out that since the S2N concerns China's development, it needs to be given comprehensive and scientific consideration. In response to this proclamation, from May 18 to June 22, 1999, eleven experts from government agencies, universities, and local research institutes organized the first examination of the proposal. Although the results of this discussion were not aired or circulated publicly, related reports suggested that this panel offered some support for the project.

At about the same time, forty-three members of the Chinese Academy of Engineering and Chinese Academy of Sciences and three hundred other experts completed the *Strategic Research on Water Resources for China's Sustainable Development in the 21st Century*. While reporting to the State Council and relevant ministries on the research team's conclusions, Qian Zhengying, a member of the Academy of Engineering, pointed out that the Western Line would not be technologically feasible for the foreseeable future. Another member of the Academy of Engineering, Pan Jiazheng, sharply criticized the WLP, emphasizing that human beings should not always try to conquer or change nature. Doing so in the past has produced heavy losses of energy, efficiency, environmental treasures, and worst of all, life. He argued that we should learn to follow the rules of nature and live harmoniously within its bounds. In his opinion, ideas like diverting water from Tibet to Xinjiang and carving the Himalayas sounded more like science fiction than engineering plans.[10]

[10] Pan Jiazheng, "Brief notes on water shortages in the North and the South-to-North Water Diversion Project," *China Water Network*, June 14, 2001.

Debate came to a head in 2006. Minister of Water Resources Wang Shucheng asserted in a speech in Hong Kong that the Great Western Line was "unnecessary, unscientific, and impractical." This would indicate that for now, the project rests on a shaky political foundation.[11]

C. *The Three Gorges Dam Project*

1. *The Impact of Water Storage* While the environmental impacts of the WLP remain possibilities, the environmental impacts of the Three Gorges Dam have become reality.

The Three Gorges Dam cleared a major hurdle on May 20, 2006, with the dam's completion. On June 6, crews demolished the third-phase cofferdam upriver, and the dam started to generate electric power. By October 27, the reservoir's depth has reached 156 meters, completing the third-phase of water storage. When complete, the water level will reach 175 meters.

The Three Gorges Dam has long received plenty of public attention because of its huge scale and influence. Speculation regarding the dam's impact went up a notch on October 27, when a rare earthquake measuring 4.7 on the Richter struck Suizhou, about two hundred kilometers northeast of the dam in Hubei Province.

Other strange occurrences in the region kept the dam in the news. From September to November, Shanghai suffered frequent encroachment of salt water into aquifers.[12] And throughout the summer, Sichuan Province and Chongqing Municipality suffered through the most severe heat wave and drought in fifty years. Many people speculated that the dam might have played a role in any or all of these phenomena.

Although the dam has not been definitively linked to any of these problems, many believe that the possibility cannot be ruled out. For instance, by adding 10.5 billion cubic meters of stored water between September and October, 2006, or by cutting river flow by two hundred and eighty million cubic meters per day, salt water has been able to flow upriver more easily.[13]

[11] China Economic Net, October 26, 2006, http://www.ce.cn.

[12] Translator's note: Bear in mind, though, that ocean tides periodically flow into the Yangtze River at its mouth and force their way upriver a few kilometers.

[13] Yu Dawei, "Salt tides to intensify in the West Lake," *Finance and Economics Magazine*, November 2006.

2. *Environmental Fallout* Upon completing the second phase of water storage, the dam and its reservoir began affecting the environment. Some major problems include:

a. *Sedimentation* According to the model tested in the demonstration phase, Chongqing's harbor will have a severe sedimentation[14] problem in twenty years. Moreover, the rising river bed upstream from the reservoir will intensify flooding in the Yangtze's upper reaches. In fact, both the Chongqing municipal government and the China Three Gorges Project Corporation (CTGPC) have made a contingency plan for dealing with this eventuality by building a new harbor. Slated to be complete in 2010 at a cost of 1.5 billion RMB, the new Cuntan Harbor will replace the Jiulongpo Harbor, which is expected to be inundated with silt. The CTGPC has proposed to use some of the profits earned from power generation to dredge the silt from the new harbor.

b. *Intensifying Geological Instability* The banks and cliffs around the Three Gorges Reservoir are prone to landslides. There have been more than 2,490 landslides and more than ninety mudslides in the reservoir area, and activity has picked up as the waters have risen. Once the water reached 135 m, several significant geological events occurred, such as the Huanglashi and Shuping landslides. Cracks, significant displacement of rocks, and massive rock distortion have been identified. Ships navigating in this section of the river often receive warnings about falling rock hazards.

Past and ongoing construction in the area and rising water levels aggravate this condition. To offset these impacts, the central government has already appropriated more than ten billion RMB to stabilize the rocks, banks, and shoreline around the reservoir area, but this effort will not completely resolve the problem.

c. *Multiple impacts on the river course, shore, and banks* As water levels have increased, the river bed has become more removed from the water's surface, while water has climbed far up the gorges. Erosion has damaged the banks already, and this condition seems likely to grow worse, even extending as much as 800 km downriver. Since 2002, rockslides and

---

[14] Translator's note: Some Mandarin sources refer to this problem as siltation.

avalanches have grown increasingly common along the Jingjiang River section of the reservoir region, as well as in the Yueyang section.

d. *Reservoir-Induced Earthquakes* After the water's depth crossed the 135 m mark, the number of small-scale earthquakes in the area increased. For example, there have been dozens of low-intensity earthquakes along the Gaoqiao fault near Badong; the strongest was a 3.4 temblor.

Note that ground movement can also occur because collapsing lime caves rapidly release gas pockets. If severe enough, this will cause measurable movement in the Earth's crust in a given locality. Large reservoirs can trigger these sorts of geological events. Such little earthquakes often happen in the Wu Gorge area, which has many lime caves, and along the Xiangxi River, where there are many mineshafts. These have produced jolts strong enough to cause severe cracks in houses located in these areas. Although low in magnitude, these disturbances may cause serious damage to structures because their epicenters are usually not very deep.

e. *Water Pollution* With completion of the dam, the river's flow will change. Former rapids are now part of a tranquil lake. This change has dramatically reduced the river's capacity to purify itself. At the same time, cities and towns now dump large amounts of untreated sewage and solid waste into the reservoir. To counter this problem, the central government has appropriated more than 5 billion RMB to build several sewage and solid waste treatment facilities in the area. However, these facilities will not come online for a few years; in the meantime pollution worsens. Water quality in the Yangtze River below the reservoir is very bad, and substantial percentages of it are classified as Level V or Level V-, which simply means that such water is badly fouled or toxic.[15] Eutrophication is common as well.

[15] The Surface Water Quality Standard passed by SEPA in 1988 rates water in six categories. Level I is suitable for river source water in nature preserves. Level II is suitable for drinking water without treatment, swimming, Grade I Protected Areas, and rare and sensitive aquatic life. Level III is suitable for drinking water with some treatment, swimming, Grade II Protected Areas, and ordinary and unprotected aquatic life. Level IV is suitable for non-body contact recreational use and industrial use. Level V is suitable for agricultural and landscaping. Below Level V (Level V-) is completely toxic and unusable. Note that Grade I protected areas include river source areas and national-level nature preserves.

f. *The Ruined Natural Landscape of the Three Gorges* At its widest, the reservoir reached far beyond the banks of the Yangtze River's pre-dam span. Notably, tributaries feeding into the Yangtze and the Three Gorges Reservoir have been widened by backwater and have become bays; roiling rapids have become tranquil lakes. This dramatically changes the original landscape, robbing the Three Gorges area of its legendary beauty. This will adversely impact tourism in the area.

g. *Serious Ecological Threats* The 600 km river section from the dam to Chongqing will be replaced by a reservoir. As a result, many fish species will lose their habitat. Fish species that have adapted to living in rapids and shallow waters will not survive in the tranquil, deep, and in some cases colder waters of the reservoir. While trying to migrate to more fitting habitats, they will likely find themselves trapped. Consequently, several indigenous species may face extinction, or at best a dramatic decline in populations. Species accustomed to swimming this stretch of river have no way to swim below or above the dam if trapped on either side of it. This can disrupt spawning cycles, and therefore put downward pressure on a given species' population.

h. *More Common and Costly Migrations* Planning for the displacement of villages, towns, and cities to make way for the reservoir depended on some largely unknown calculations—specifically, estimates of how the hydroelectric gradient, and rate and degree of sedimentation, would affect the final water level of the reservoir, and thus the number of people evacuated.[16]

In October 2006, the Xinhua News Agency cited Pu Haiqing, the director of the Office of the State Council's Three Gorges Project Construction Committee, as having said that the total number of displaced persons would be 1.4 million, almost twice the 725,500 estimated in the

[16] Translator's note: Before building the dam, engineers calculated the gradient in the Yichang-Chongqing section of the Yangtze River as 2.3/10,000 degrees, meaning that the river rose only twenty-three meters for every 100 km of river. During the dam's planning stage, the team in charge of migration issues thought that the gradient could be ignored because the dam is 493.9 km from Lidu County in Fulin, but the difference in water level is only 0.4 meters. However, the team in charge of sedimentation held that the number should be 7/10,000 degrees. After filling the reservoir, the CTGPC and the Chongqing Bureau of Migration told the press that the average waterpower gradient is about 0.53/10,000 degrees. According to this calculation, the reservoir's waters will rise as high as 177 m, flooding many more homes than originally estimated.

official 1992 report submitted to the National People's Council (NPC). Note as well that as the instability of shoreline areas increases, and as silt levels rise, more people will need to leave the area. The final number of people likely to move has yet to be seriously considered.

i. *Insufficient Lock Capacity* During the construction of the Three Gorges Project, overland routes were used for transportation typically conducted in the Yangtze River. Although originally intended as a temporary measure, overland traffic has remained high because of bottlenecks in ship traffic through the Three Gorges Locks.[17] The severity of the problem can be understood by examining data from the second half of 2003, since the locks opened in June 2003. Ships became backlogged on 119 of the 197 days that locks operated in 2003, according to the Three Gorges Navigation Administration Bureau.

The reason for this is both simple and disappointing: the locks have not lived up to their billing. With an "official" capacity of fifty million tons, the locks can only carry thirty-four million tons in daily capacity in reality. This falls well short of the growing commercial demands of shipping interests in the area.[18]

Diverting so much traffic to overland transit vastly increases emissions and oil demand. Both of these effects have serious environmental impacts for the Three Gorges area as well as the broader ecology.

II. *The Environmental Impacts of Large-scale Construction Projects*

A. *Impacts on Geological Conditions*

According to Guo Kai's report *Build the Shuotian Canal and Revitalize China*, delivered at the Fragrant Hills Conference on July 1, 2005, the Great Western Line Project does not take into consideration the

---

[17] Translator's note: Since June 2004, container ships have been barred from passing through the locks. Instead, containers travel on overland routes. Passenger traffic has also been diverted to land routes. Even with these measures in place, the locks cannot handle the current demand.

[18] Translator's note: Note that these capacity figures do not reflect downtime in operations. Many factors could further reduce lock operations: thick fog, floods, adjusting reservoir capacity, clearing sediment from the locks, and other forms of routine maintenance and inspection.

limits and impacts of geological conditions and geological instability on construction projects. Indeed, some of the techniques proposed in the Great Western Line Plan, such as using detonations to cause landslides in order to create rock dams, need to be reconsidered. For one thing, this kind of construction will likely cause or lead to serious geological damage. Some provisions in the *National Statute on the Prevention and Control of Geological Damage* forbid the use of such construction techniques in various areas. Also, while rock and earthen dams have been used throughout China, especially in eastern Tibet and western Sichuan Province, these structures have obvious problems. They are leaky, easily toppled, and prone to erosion and collapse. For instance, an especially large landslide in Yigongzangbu in 2000 created a natural dam more than ten times larger than the Three Gorges Dam: 4.6 km in length, 3 km in width, and sixty to one hundred meters in height. It collapsed after only sixty-two days. Even in this short time, a 52.7 km^2 lake reaching depths of 62 m had formed. The wall of water released with the dam's collapse devastated downstream areas.

It remains to be seen if a landslide of this scale can be created artificially. Additionally, it is not clear how large an artificially-created earthen dam can be. These basic questions need to be addressed to better understand the impact and feasibility of the Guo Kai plan.

Neither the Western Line Plan nor the Great Western Line Plan takes into consideration the special geological conditions of the Qinghai-Tibet Plateau. Even if the plans can be constructed, ongoing operations and maintenance will be very risky and costly.

Other large construction projects in the area provide some insight into these kinds of complications. The highway that leads to Motou, Tibet, serves as an especially poignant example. Soon after its completion, it had to be shut down after an avalanche. Attempts to repair the road have ended in failure for technical and financial reasons, so now the road has been abandoned. The Bomi-Nylingchi section of the Sichuan-Tibet Highway has a reputation as the most dangerous section of the entire road. It requires ongoing maintenance and repair that has proven exceptionally costly. Worse, the road has been closed for some especially large repairs, and cannot be used during the frequent bad weather that occurs.

Road closures have been an especially common problem for the No. 102 ramp east of Tongmai. The road was closed on about ninety days each year between 1991 and 1997. This six hundred meter ramp has absorbed fifty million RMB. One proposal for resolving the problem, or at least lessening the costs and improving road access, involves building a bypass tunnel as a replacement. However, this comes with a high up-front price tag that could reach two hundred million RMB.

Maintaining either the Western Line or Great Western Line will prove even more complicated and expensive. A similar but smaller scale water diversion project makes this clear. The Three-River Water Conservancy Project in Tibet began construction in the 1980s. The project included dozens of irrigation stations and thousands of meters of aqueducts. However, avalanches frequently disrupted construction; damaged and destroyed facilities had to be replaced and repaired even as basic construction moved forward. During the 1990s, avalanches and landslides destroyed the entire project. Three billion RMB had been spent for nothing.[19]

B. *The Impact on Water Resources*

These projects will likely influence water resources in at least two major ways. First, diverting water from the arid plateau will accelerate desertification and further degrade water quality in river valleys downstream from the areas where water has been diverted. Second, slowed and reduced water flow will weaken the affected rivers' purification capacities. Pollution will worsen; the reduction in water quality will deliver more water, but less of it will be useable.

For example, the water storage capacity of forests in western Sichuan and northwestern Yunnan has decreased from four hundred billion cubic meters fifty years ago to one hundred billion cubic meters now. The arid valley of the Yalongjiang River has expanded, rising two hundred to four hundred meters higher into nearby hillsides and plateaus. The annual flow of the Daduhe River has decreased by 14 percent from its average volume over the latter half of twentieth century. Moreover, its volume fluctuates more widely, with the flow in dry seasons becoming much lower than it had been even a few decades ago. If the S2N is built according to the Guo plan, it will divert 32 to –36 of the water volume from these rivers. A number of river sections will be acutely

[19] Translator's note: The RMB trades at about 8:1 against the dollar. Therefore, this amounts to about 375 million dollars.

disrupted: the 900-kilometer section of the Jinshajiang River between Yushu and Panzhihua; the 600-kilometer section of the Yalongjiang River between Ganzi and Panzhihua; and the 300-meter section of Dadu River between Touerduo and Shimian. Some sections near the dam may be completely blocked. Taken together, these changes will expand the arid valley and exacerbate desertification and grass degeneration on the plateau.[20]

As originally proposed, the Great Western Line Project will divert two hundred billion cubic meters of water annually. In 2005, Guo Kai revised this figure downward to only 128 billion m^3, and set a minimum of one hundred billion m^3. Even with these adjustments, between 83.3 percent and 91.5 percent of the water volume of some rivers will be diverted. So for the Suma Plain Fork of the Yarlung Tsangpo River, diversion would redirect thirty billion m^3 of the river's annual 32.8 billion m^3 flow. For the Shangbaiba section of the Niyang River, fifteen billion of 16.5 billion m^3 would be diverted; the Shangbagai section of the Yigong Zangbu River would lose fifteen billion of its eighteen billion m^3 annual flow; Shuowaba Canyon of Nujiang River would lose thirty of 35.6 billion m^3; the Changdu Dunda branch of the Lancang River would see twenty of twenty-three billion m^3 redirected. In the Western Course of S2N, the flow to the Tongtianher, Yalong, and Daduhe Rivers would drop by 69.4 percent.[21]

Rivers must maintain a minimum flow rate in order to maintain their basic functions. These functions are multidimensional: maintaining water purity and self-cleansing; supplying water for the surrounding ecosystem, including tributaries; and providing for human economic and survival needs as well as recreation. Reducing water volume beyond the minimum flow rate harms rivers and riparian environments. Various experts agree that rivers must retain 50 to 70 percent of their annual flow in order to maintain these basic functions. Put another way, humans should limit their use to between 30 and 70 percent of a river's waters. Clearly, the aforementioned diversion projects vastly exceed these percentages.

[20] *Memo for the Western Course Project of South-to-North Water Diversion Project*, 83–84.

[21] Yellow River Conservancy Commission, Ministry of Water Resources, *The Guidelines for South-to-North Water Transfer Project and the Plan for the First Stage of the Project* (2001).

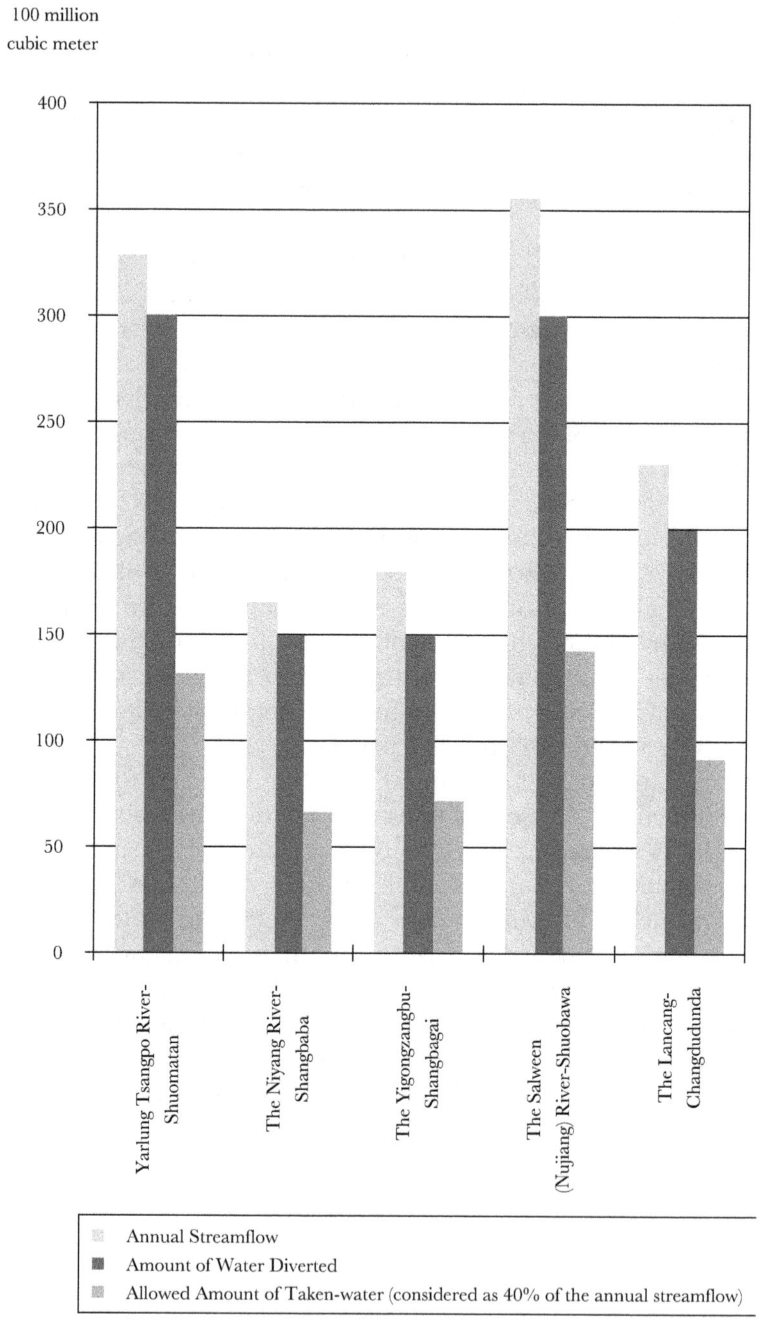

Figure 11.1: Proportions of water diverted by the Great Western Line Project

The Great Western Line Project includes plans to make Qinghai Lake a reservoir, increasing its volume from eighty billion m^3 to three hundred billion m^3. Water resources minister Wang Shucheng considered this project "ridiculous" since Qinghai Lake is a salt lake. Guo countered that by increasing the lake's volume by 275 percent, salinity would decline from 14 grams per liter to 1 gram per liter. Of course, this does not address the impact on the local ecosystem, particularly on species dependent on high-saline water.

C. *Environmental Influences*

Both the Western Line and the Great Western Line schemes impact mountainous southern Tibet, the alpine and valley regions of western Sichuan and eastern Tibet, and the plateaus and broad valleys of southern Qinghai. These areas include prairies, meadows, and forest ecosystems. As the source area of the major rivers sustaining China, Southeast Asia, and parts of India, the region is at once precious and fragile. Therefore human activity can cause rapid and deleterious harm. Erosion of glaciers, declining water levels in lakes, worsening soil erosion, widespread desertification, and grassland deterioration stand out as some of the most obvious threats to the highland ecosystem.

The ecosystems of the mountainous areas in southern Tibet, northwestern Sichuan, and southern Qinghai have been affected more than other places. The decline of the Ruoergai Wetlands is a typical example. The enormous construction work necessary for the water transfer projects will further damage and destroy vegetation and cause profound changes in the surface environment, biological environment, and general ecosystem due to excessive excavation, exploitation of sand stones, and the disposal of refuse and waste in connection with large infrastructure projects.

The Great Western Line will traverse the mountains and canyons in eastern Tibet, cutting through some of China's last-remaining, fully-intact old-growth forests. The area is especially rich in biodiversity, representing 60 to 70 percent of all Tibetan species. Construction would disrupt life for many of these species and the integrity of the forest biome.

None of this has been reflected in official reports or books describing the Great Western Line Project. Planners simply overlooked the impact on natural preserves, protected forests, biologically diverse areas, and historic treasures as if they did not exist. When asked about this, Guo Kai unapologetically stated that, "Nowadays peat moss[22] fetches a high price on international markets. As part of the construction, we can dig it up and sell it to bring in foreign exchange." This kind of callousness and insensitivity is stunning, especially considering how much forethought went into developing this plan.

D. *Impacts on the Social Environment*

The S2N Project will draw upon the slower courses of the Dadu and Yalong Rivers, and the upper reaches of the Jinshang River, for its water supply. Tibetan villages, farms, and temples dot this landscape. Dams planned for this area will create reservoirs that will displace thousands. Because of their rich cultural customs and values, especially their deep connection to the land, relocating these persons, and their communities, will prove exceedingly difficult.

Moving the Tibetan people is different from moving thousands of Tibetans. For thousands of years, these people have considered the Qinghai-Tibet Plateau to be a Holy Land. The divine mountains and lakes are the carriers of Tibetan culture. The admiration and worship of these mountains and lakes have become fixed rituals; by Tibetan tradition they cannot be altered, used, or harmed. In essence, these are physical and spiritual presences. These values have been central in protecting Plateau ecosystems. More than this, though, the Tibetan way of life completely belongs to and within the Plateau's ecology.

As it stands, the proposed water diversion projects will significantly alter mountains, lakes, rivers, forests, and grasslands considered holy by the Tibetan people. Temples and shrines fair no better: plans for their razing, reconstruction, or removal have been included in project plans. Obviously, such violence to land and culture will greatly impact the people who live and work in these places, as well as the extended community of Tibetan Buddhists. Conflict over these plans could intensify and become a serious problem for national unity and safety.

[22] Translator's note: Peat moss is sometimes called sphagnum in Mandarin-to-English translations.

The Great Western Line Project proposal has failed to execute any serious analysis of the plan's social and cultural impacts. For instance, if the ninety-meter-high Shuomatan Dam, at the starting point of the Great Western Line Project, is built, it will raise water levels by eighty meters. This will submerge much of the river valley areas of the Yarlung Tsangpo River from Zedang to Qushui. Yet some experts claim that these are the most prosperous regions in Tibet.

Guo has gone so far as to claim that the dam would produce only negligible impacts. In at least one version of his plan, cultural hubs like Changdu and Luhuo, and numerous famous temples, would be submerged. Eventually Guo relented after a 1999 project review, and alterations have been made.

Creating reservoirs on the highland will also worsen economic conditions for the Tibetans. Grassland can no longer keep pace with the demands of local livestock production.[23] Storing more water in the region, ironically, will worsen this condition, since rats will likely multiply and end up ravaging what little grass remains. These problems require serious deliberation before the project moves forward. Unsurprisingly, plans have completely neglected such issues. This huge omission seems likely to create major problems later if the project is ever realized.

E. *International Impacts*

Some of the rivers involved in the Great Western Line Project, like the Yarlung Tsangpo, the Nujiang, and the Lancang rivers, flow across international borders. No universal agreement on international rivers exists; nations sharing a river course or river basin must fall back on two, often conflicting, principles: national sovereignty and international cooperation. The former allows a given nation state to use river water any way it wishes; the latter stipulates that the development of water resources should not impair other nations' interests. Note, however, that calls for cooperation lack enforcement mechanisms, which privileges sovereignty over cooperation.

---

[23] Translator's note: The problem is that local livestock production now supplies regional and national demands as well as the relatively small Tibetan and highland population.

The 1972 UN *Declaration on the Human Environment* clearly states that:[24]

> States have, in accordance with the Charter of the United Nations and the principles of international law, the sovereign right to exploit their own resources pursuant to their own environmental policies, and the responsibility to ensure that activities within their jurisdiction or control do not cause damage to the environment of other States or of areas beyond the limits of national jurisdiction.[25]

The 1992 *Rio Declaration on the Environment and Development* reiterated this principle while emphasizing development, especially within a cooperative framework.[26]

Clearly, projects as large and influential as those proposed for diverting water on the Tibetan Plateau and surrounding areas have the potential to set off international disputes. These projects affect ecosystems and food supplies. For instance, proposed changes could produce dramatic declines in fish populations or in water volume needed for agriculture or aquaculture, and could create water access problems. Floods could result from diversion projects if any part of these systems failed. Behind all of the analysis are the fundamental issues of life or death, and not just for people; other species and the health of ecosystems may hang in the balance.

Such problems are not idle speculation. After the great landslide in Yigong Zangbu in 2000, the Chinese government warned countries downriver, including India and Bangladesh, of impending floods. Despite this, however, the flood killed many persons and destroyed vast amounts of property.

Elaborate plans for multiple-dam projects have raised objections from downriver countries. Given their grandiose scale, it's no wonder. One such plan would put a cascade of eight dams with two reservoirs on the Lancang River around Xiaowan and Nuozhadu; another would include a staggering thirteen dams on the Nujiang River. Needless to say, the scale and impact of these projects would have serious repercussions for south and southeast Asian countries that rely on these rivers for myriad uses.

---

[24] Translator's note: The reader should note, however, that the *Declaration* is a non-binding agreement.

[25] Translator's note: See http://www.unep.org/Documents.Multilingual/Default.asp?DocumentID=97&ArticleID=1503 for the complete declaration.

[26] Translator's note: The reader should note that the Rio agreement was binding.

Even dams in downriver locations have produced serious environmental impacts. The Menghe Dam, built by Thailand on the Mekong River, blocks fish migration routes. The number of fish species in this stretch of the river has fallen dramatically, from 265 to ninety-six. The annual catch in the region has fallen by 80 percent.[27] This has a profound effect on those living in the area. Fish supplies provide livelihoods and better, healthier lives. Ninety percent of the persons living in the Mekong basin farm; 80 percent of their protein intake comes from freshwater fish. The annual catch amounted to one hundred million tons, and brought in over one billion RMB before the dam was built.

The Times of India reported on the Great Western Line Project in 2006. These and other reports indicate that the Indian government has discussed the project at some length. Needless to say, the project has caused great concern; some fear that crucial water supplies could drop by 50 percent or more. The Indian government plans to engage the Chinese government in high level discussions about these projects.[28]

III. *Preventing Problems Associated with Large-scale Projects*

A. *Considering Sustainably from Start—to Finish*

Huge projects like the South-to-North Water Diversion Western Line and Great Western Line have significant, wide-ranging, and sometimes surprising environmental impacts. Starting these projects commits policymakers, and those affected by policies, to the results of these actions. Careful forethought and comprehensive planning play pivotal roles in organizing and implementing large-scale projects.

Time and time again, project supporters claim that water shortages in northern China necessitate building these water transfer systems. Certainly, water supplies in the North run distressingly short. Still, these projects do not make logical or economic sense. This project rests on the misbegotten notion that man can, and perhaps should, conquer Nature. In fact, Nature is considered little more than an obstacle or

---

[27] Wang Xiaomin, "The Environmental Influence of Hydroelectric Development on the Mekong River," *Southeast Asian Studies* 3 (2001): 13.

[28] Tao Ye, ed., China Economic Net, October 23, 2006, http://www.ce.cn.

an afterthought, as noted by their repeated oversights of ecological impacts. From the standpoint of men like Guo, Man plays the central role in earthly matters. Projects become monuments of ego: the bigger, the better.

In a broad sense, these projects may well fail to pass economic muster. However, they can bring substantial money to particular contractors, project managers, investors, and their cronies. Project boosters claim that benefits are widespread; even though some persons in Sichuan may suffer somewhat, the overall benefit outweighs their collective losses. Moreover, the claim goes, those who suffer can be compensated. This perspective overlooks many aspects of project effectiveness and economic rationality.

First, this system tends to raise supply to meet demand. It undermines conservation efforts, which might well be less costly. China continues to use water very inefficiently compared to other countries. This means that better water management could bring supply much closer to metting demand, significantly reducing shortages. Certainly, spending on conservation efforts would save billions, rather than squandering money on grandiose water diversion schemes. Second, overestimates of the effectiveness of diverting water call the entire cost/benefit analysis into question. As noted above, estimates of total water volume in the project keep falling; external estimates suggest that estimates of total water volume diverted cannot be met. Given the amounts that can be delivered, the project serves as a stopgap measure at best. Demand will once again outstrip supplies within fifty years.

The high cost of the project puts water costs as high as twenty RMB per m^3; desalinization is expect to drop to about three RMB m^3 within twenty years, according to Minister of Water Resources, Wang Shucheng. Investing now in desalinization research and the infrastructure needed to operate and maintain such a system seems like a wiser and more durable investment.

Remarkably, some drought-stricken areas in northern China lie in sub-humid zones. Given this, we see that human activity drives shortages, not a lack of water. For example, excessive cultivation, damming, and irrigation have caused many lakes in Xinjiang Province to run dry, including Luobupo Lake and Black Lake. Water shortages affecting Tianjin have their roots in the inappropriate use of the Haihe River, particularly in a series of canals that link the river's tributaries into a navigable network.

All this bluster about diverting water obscures the deeper and more important question: "What is the best way to resolve water shortages for the least cost and greatest overall long-term benefit?"

B. *Choice and Livelihood*

Boosters promote large-scale projects by claiming that they reduce poverty and facilitate development. The facts do not support this view: benefits tend to concentrate in the hands of developers, contractors, and local government officials. Few others see tangible benefits; many suffer in the wake of such projects.

Land development and use mechanisms drive this process. In what amounts to a system of imminent domain without adequate compensation, commercial land development interests can buy—or nearly seize—collectively-held land from farmers for bargain-basement prices. This dramatically lowers operating costs. When revenues come in, a greater and earlier return on investment is realized. This fattens corporate bottom lines, but multiplies the number of persons with little-to-no wealth holdings or the means to make a living. Without receiving adequate compensation, those displaced descend further into poverty. Many people can be adversely affected by a single transaction, while only a few will ultimately benefit. By widening the gap in wealth, such moves intensify social tension.

While this is all legal, it smacks of market-based predation. Local and provincial governments, alongside developers, agree on land prices; those occupying and working the land have little if any say about the price of land. Government, which clearly has a financial interest in striking bargains with developers and corporations, also enforces the transfer of assets and resettlement of the former occupants. Government and developers expect migrants, farmers, and those with ancestral holdings to simply acquiesce.

For example, the Danjiangkou Reservoir of the S2N Midline section displaced 412,000 people. After unsuccessfully working the land in the resettlement area, 108,000 of the evacuees retuned home.[29] However, with much of the old land inundated with water, these persons had

[20] Wang Maofu, et al., "The Push and Pull Role the Environment Plays in China's Migration," *China's Demography* 3 (2000).

to pack onto smaller plots and farm in tough and rather inhospitable conditions. Land holdings fell to only 0.4 *mu* (1 acre = 6.07 *mu* or 1/15 of a hectare) per capita. Worse, farmers began to work the nearby slopes; some of these are pitched between twenty-five and forty-five degrees. More than ten thousand of the returnees eventually had to move again as the reservoir continued rising.[30]

Those displaced had been asked to make a personal sacrifice for the greater national good. This argument runs counter to law and migrant rights policies. Asking these persons—many of them poor Tibetans—to sacrifice for the good of others, including those who tend to be better off in material terms, worsens social inequality and increases socioeconomic tensions. This conflicts with the goal of building a harmonious society.[31] It also undermines the smooth functioning of capitalist markets by trampling on property rights, and using state power to transfer wealth based on political access and clout.

The following example paints a vivid picture of these forces in action. The China Three Gorges Travel Agency, under the head office of the CTGPC, developed its operations with tax support and public savings. After monopolizing tourism in the Three Gorges area, it started charging visitors high ticket provinces to gain access to the dam site. Not only did they limit access to a public site built with public funds, their monopoly operation squeezed out local operators, including those run by evacuees.[32]

Projects adversely affecting minority peoples should be redesigned. Cultural and social conventions and legacies need to be considered within the planning context. Furthermore, natural conditions, and their connection to local culture, need to be better understood and integrated into project planning and environmental impact assessments (EIAs). Narrow-minded and misguided thinking related to these points, such as ethnic prejudice and a "Han-centric" sense of cultural superiority, should be set aside. Some Han people, besotted with their own ideas and desires, seize land and resources in the country's western region

---

[30] The Reservoir and Resettlement Bureau of the Ministry of Water Resources, *Briefings of Resettlement Undertakings* 1 (2003).

[31] Translator's note: Aiming to build a "harmonious society" is a basic tenet of Chinese social policy. This multidimensional goal seeks to create balance, cooperation, and coordination between different parts of society, among many other objectives.

[32] Gui Huiqiao, "The Three Gorge Dam Becomes the Source of Money for Some People," October 1, 2006.

as if they owned them. This ignores the longstanding management systems in place in the West, violating cultural and individual rights of persons and groups living in minority regions.

C. *Dispelling the Man versus Nature View*

Proponents of the Great Western Line Project claim that it will create a new China, joined by a massive water distribution network that turns desert into fertile farmland. Going beyond the bravado, we hear echoes of the Great Leap Forward's rhetoric, which claimed that "man can always change and conquer Nature." However, the science simply undercuts such arguments.

Managing a fixed resource should encourage a given community to cooperate with its neighbors and the surrounding ecosystem. The western region certainly faces stringent water constraints. With limited, and falling, ice and snow melt, oases and rivers play a special role in the ecology of the deserts and wilderness of the region. However, artificial intervention—creating more numerous and enlarged water sources, for instance—can expand use and population beyond capacity constraints. Eventually, this can lead to overtaxed or degraded manmade systems, as wells, or large water transfer projects, become more stressed and vulnerable. Tampering with resource supplies to meet demands eventually runs up against biological, ecological, and physical capacity constraints. We cannot arbitrarily distribute natural resources and overhaul the natural environment to meet human, industrial, economic, and political needs at will without paying a high cost for such hubris.

D. *Developing Sound Scientific Evaluation*

Within a centrally-planned economy, political leaders with power and political capital develop and execute strategic decisions regarding large-scale projects by fiat. Note that these individuals and the organizations they work within tend to begin decision making about such moves with a great many things decided beforehand. Most fundamentally, undertaking the project may be a *fait accompli* by the time any kind of EIA or other scientific assessment begins. EIAs even include such obscure political code as "the project has no fatal flaw (and should go forward)" or that "negative impacts will be mitigated through various measures once the project is underway." Remarkably, this gives the green light to a *flawed* plan even though it is not *fatally* flawed. In other cases, analysis orbits around gathering minutiae in defense of a given plan; the large of question—should we do this or not?—is often overlooked or avoided. Such rhetorical flourishes obscure discussion of real environmental impacts.

However, such rhetoric seems to work effectively at building support for these projects. For instance, online debate about the Great Western Line Project sometimes defends the idea based on its being heartfelt and audacious. One commentator on the Internet countered with this argument:

> In a society where scientific, rational, and effective decision-making dominates, putting forward more ideas expands the marketplace of ideas. On the other hand, within a mechanism that works scientifically and effectively, it is always a good thing for ideas to be put forward whether they themselves are scientific or not, because this at least gives decision makers more options. Yet, with a less thoroughly defined decision-making mechanism, more ideas can produce more temptations to put them into effect. Audacity or strength of conviction may carry the day rather than sound, sober thinking. Acting on such impulses not only puts immediate interests at risk, future generations may have to bear the ill effects of such moves. Too many mistakes like these have been a part of our history.

Tibet's Water Saves China fits this framework as well: this is literature with a clear polemical thrust. It excites the passions of its readers; it reaches deep into the popular consciousness. This sense of enthusiasm leads many readers to believe that the proposals are feasible. Building popular opinion begins to get the wheels of "progress" moving. However, readers and the public should not confuse this book's arguments with scientific analysis.

Returning to the case of the Three Gorges Project provides a useful illustration of this sort of manipulation. Project plans called for the water level to rise from 135 to 156 meters over a four-year period concluding in 2007. However, accelerating this process would increase electricity output, and therefore, revenues. Needless to say, the economic incentives overtook advanced "planning." The reservoir filled to 156 meters one year ahead of schedule.

A follow-up plan projected water to rise again, to 175 meters, from 2007–2013. Now project managers hope to meet this goal by 2009. These revisions have been introduced, and in some cases adopted, without any scientific analysis. They have also flown in the face of advice offered by many experts calling for managers to stick with the original plan in order to minimize environmental impacts. It remains to be seen what the final impact of these hasty and haphazard decisions will be. But implementing these sorts of mechanisms, especially in large projects, can have serious unintended environmental effects.

APPENDIX

HAS THE THREE GORGES DAM CAUSED THE DROUGHT IN CHONGQING?

In the summer of 2006, Chongqing and Sichuan Provinces suffered unusually high temperatures during a severe drought. The temperatures in Chongqing reached 44.5 degrees Celsius, a fifty-three-year high.

This drought does not exist in isolation; this is but one manifestation of a larger systemic change. First, global warming may well be at work in this case. Second, and more directly, experts have linked the drought to anomalous activities of Pacific subtropical and Tibetan high-pressure systems.

In the winter of 2005 and the spring of 2006, the ocean surface temperatures rose by 0.5°C above average. A subtropical high pressure system had moved northwest, blocking warm, moist air currents from the eastern part of China. The western Pacific high-pressure system had been disrupted; wind activity and cloud covering dropped off, leading to higher temperatures in the region. The western Pacific subtropical high-pressure system moved west of its usual position, which acutely affected Chongqing and the Sichuan basin.

While temperatures rose, water supplies declined. The amount of snowfall on the Qinghai-Tibet Plateau fell 20 percent short of its average, reducing the area of the snowpack by 10 percent from yearly averages. The number of snowy days fell ten and thirty days short of averages, depending on the specific location on the Plateau. Unsurprisingly average temperatures hovered 2 to 4°C above average. Thermal effects produced less snow, and when it did fall it melted more readily. Ambient air temperatures rose as a result. This warmer air held more energy and more moisture, energizing the Asian monsoon. With more energy, the monsoon shifted northward, depriving the Yangtze River Basin of its normal precipitation.

However, some have implicated the Three Gorges Dam in some part of these changes. Wang Hongqi, a scholar in geography based in Beijing, sees a direct link. Wang uses a metaphor to make his argument: if the Sichuan Basin is like a big cask and the Three Gorges is its shortest board, the naturally narrow gap channels air flow into the basin. The Three Gorges Dam lengthens the board, substantially reducing water vapor circulation. With this, air temperatures will rise. He notes that meteorologists, by focusing exclusively on middle- and high-altitude air currents, ignore the impact of lower altitude air currents (below one thousand meters) on climate.

However, Dong Wenjie and Wang Qiwei from the National Meteorological Center find Wang's argument untenable. Because the dam stands less than two hundred meters tall, it cannot affect the major vapor and heat currents, which flow through a high altitude channel several thousand meters above the surface rather than through the narrow Three Gorges Valley channel. Domestic and international experts corroborate their view and assign a negligible weather impact to the project, stating that "The reservoir can only affect the microclimate of an area a few kilometers around the reservoir. It is not capable of affecting the climate of an area as large as Sichuan Province."

Wang Hongqi agrees that the project is not the major driver of weather change in the region, but sees manmade factors, including the Three Gorges Dam and Reservoir, as having significant impacts.

Wang grounds his argument in a systemic analysis of water-vapor and air-flow patterns in the region. Water vapor from the Indian Ocean supplies much of the Sichuan Basin's supply. These vapors flow through the valleys in the Hengduan Mountains. Upon arriving in the Qinghai-Tibet Plateau, it turns to rain or snow, which feeds the Yangtze River. However, in recent years dozens of dams have been built in various valleys along the Jinsha River and Lancang River, both major offshoots of the Yangtze. These dams block the northbound movement of water vapor from the Indian Ocean, which reduces precipitation on the Qinghai-Tibet Plateau. Furthering his point, Wang holds that these disruptions influence, at least in part, the anomalous activities of Tibetan high-pressure systems.

Wang believes we need to look beyond narrowly-focused explanations for the drought, such as odd behavior in high-pressure systems and unusually sparse snowfall. He advocates undertaking a research program that looks more deeply into long-term trends, such as the multiyear decline in snowfall on the Qinghai-Tibet Plateau.

CHAPTER TWELVE

ARE FENCES AND GRAZING BANS THE BEST TOOLS FOR CONTROLLING DESERTIFICATION?

Liu Shurun

Recently, the Chinese government has put considerable emphasis on controlling sandstorms. A favorite measure applied repeatedly in these efforts has been to fence off large areas of grassland. After initiating these moves, sandstorm activity declined, lending credence to the measure's effectiveness. Plans call for the amount of grassland to be enclosed by fencing to expand. However, in 2006 many sandstorms, more fierce than before, swept across the north and northwest. These storms have cast doubts on the effectiveness of the enclosure strategy, and critics of the approach have re-emerged. They not only question the effectiveness of using fences to protect grassland, they have deep misgivings regarding the impact of the enclosure program on local cultures and economies based on herding throughout grassland ranges.

The author suggests that preventing desertification goes beyond pure science; it concerns people's beliefs, perceptions, and ways of understanding connections to land and place.

Overgrazing can degrade the grassland ecology, leading to desertification. Therefore, fencing off grassland and banning grazing seem like straightforward approaches to counter this result. However, working in the field with herders provides insights that call into question this linear approach to "solving" the problem of desertification. Ecologists maintain that traditional methods of grazing and grassland use are at fault, but their narrow views fail to effectively resolve problems of grassland overuse and desertification while protecting useful cultural traditions. This article goes beyond the standard models and misunderstandings to uncover new, and hopefully useful, perspectives in protecting grasslands and grassland cultures.

Key words: sandstorm, fencing off rangeland, grazing ban, desertification

I. *Grazing Bans Are Unnatural*

Rehabilitating vegetation is the single most potent means to prevent sandstorms. Many view grazing bans through grassland enclosures as the most effective, least expensive, and practical means for rehabilitating vegetation. Examples of the direct and measurable impact of fencing off grassland abound.

The village of Hunshandake lies on a grassy plain in the Blue Banner[1] of the Xiliguole League.[2] The plain largely consists of two environmental types: *Tuo* and *Dian. Tuo* refers to sand dunes; *Dian* refers to willow bushland, marshes, and all kinds of meadows. Typical grass harvests may reach five hundred *jin* (half a kilogram) of grasses per *mu* (one-fifteenth of a hectare or just under one-sixth of an acre). This is relatively high productivity compared to other grassland. Over time, however, yields have declined, falling to only 10 percent of the former output. To halt the decline, the area was fenced off. After a year, the grass had grown to a height of 80 to 140 cm; output rose to five thousand one hundred jin per *mu* in 2001. In 2003, grasses reached 185 centimeters in height. Output attained a new record of six thousand five hundred *jin* per *mu*, or about 100 times grass output in areas where desertification had set in.[3]

Although fencing off grassland may seem to return grassland to a natural state, the fact is that enclosing land introduces human intervention into natural processes. With grass-eating ungulates totally excluded from grasslands, grass grows higher than would ordinarily be the case. In this particular area, several factors drive this process: a rich soil produced by layers of manure being impressed into the soil; ready access to good water; and favorable wind conditions. Given these factors, the enclosed area produced grass at above-normal rates.

Excluding grass-eating mammals reorganizes the ecosystem: new manure supplies do not fertilize the soil; grasses grow unnaturally high; and the turning of the soil produced by animal hooves does not take place. In short, what at first blush seems like a return to a natural state turns out to be a move toward a new kind of unnatural state.

Note as well that the grass production figures mentioned above depend on harvested grass, not the total amount of grass grown. Grass used after harvests, as well as grass used as feed, do not appear in output

---

[1] Translator's note: Banners are essentially county-level government administrative regions in Inner Mongolia.

[2] Translator's note: Leagues are essentially prefecture-level government administrative regions in Inner Mongolia.

[3] Jiang Gaoming, "The restoration of zoology: we should believe in the force of nature," in *Crisis and breakthrough of China's Environment*, Liang Congjie and Yang Dongping, eds. (Beijing: Social Sciences Academic Press, 2005). Also available in English as *The China Environment Yearbook (2005): Crisis and Breakthrough of China's Environment* (Leiden: Brill, 2007).

figures. In short, the post-ban figures artificially inflate productivity, and therefore, the benefit of grazing bans.

Finally, the enclosed areas are further divided by wire netting. This impedes movement across the plain, dividing the plains into a patchwork that is difficult to navigate. The whole system runs counter to natural processes and the culture of the nomads. Furthermore, building and maintaining the netting absorbs considerable time and money.

II. *Can Fences Lead to a Rapid Restoration of Grassland?*

Fencing off grassland can achieve solid short-term results, but will not provide a comprehensive, long-term solution to grassland degradation and desertification. Even the short-term results may be fleeting. More worrisome, measurement of "recovery" relies far more on quantitative results than any qualitative assessment of these results. For instance, a considerable portion of the "harvest" consists of weeds and invasive species; the original grasses have not fully recovered. The enclosure area maintains a degree of stability, but does not resemble the dynamic, open-ended ecology that existed before.

Many questions remain concerning desertification and grassland restoration. Cataloging some of the symptoms of desertification helps to clarify what happens in the process. The process takes hold in areas where indigenous vegetation has declined, lands have been taxed through agricultural mismanagement and overgrazing, grasslands have become sandier, and salt marshes have appeared. In other words, both natural and manmade forces can expand deserts.

Indigenous plant species in deserts and semi-deserts usually grow slowly and have a low maximum height. However, they are well adapted to local conditions. They help to stabilize land and ecosystems, and therefore are crucial to the continued health of these areas. However, because these plants do not grow to be very tall or dense, they produce a low cover rate by some standards. Solely relying on the cover rate, or harvest size, as signs of health in grasslands overlooks natural ecological tendencies.

In the enclose-and-ban system, high output grasses are favored over mixed plant groupings. This logic has led to some especially perverse outcomes, such as the replacement of indigenous flowers, like violets, with woodland forest. Under the system of measurement used, this counts as a success because ground cover and plant-matter volume both increase. However, such tree plantings often rely on nonnative species, and further, cause disruption in local ecosystems.[4]

Various experts report that total desert area has declined between 1999 and 2005. In Inner Mongolia, the desert has relinquished 16,059 square km, or about 26 percent of the total desert area recorded in the 1990s.[5] With the number of sandstorms dropping during these years, controlling desertification seemed like a simple matter of fencing off grassland and letting "Nature" do its work to restore ground cover. Seemingly, the problem of encroaching deserts had been halted. However, Mother Nature had not been placated or tamed; the huge sandstorms of 2006 made this emphatically clear.

Deserts can expand rapidly over vast areas, and understanding their spread can prove useful. Two examples of rapid desertification make for handy case studies: the Dustbowl in the American Plains States in 1934,[6] and the still-worsening sandstorms in Central Asia, especially around the Aral Sea.[7] In the former case, the problem proved transitory; in the latter, chronic. While we can hope that grasslands in China will quickly recover, we may well be in for trouble for years to come.

Working with grasslands in China will require a long-term commitment and a more systemic approach to finding solutions. Rather than banning grazing outright, reducing grazing and integrating livestock into the ecosystem can achieve steadier and more durable results. Facts have shown that by reducing the number of livestock and adjusting

[4] Translator's note: Such plantings have also proven vulnerable. The most famous failing of a manmade planting is the Northern Shelter Forest, which has absorbed millions of RMB but has been rendered ineffective by pestilence. See "Ecological Recovery: Believing in Nature" by Jiang Gaoming in *The China Environment Yearbook (2005): Crisis and Breakthrough of China's Environment.* Liang Congjie and Yang Dongping, eds., (Leiden: Brill, 2007).

[5] Bai Jianhua, "The status quo and strategies in dealing with Chinese desertification: the control of desertification and the restoration of zoology," *The China and Korea Desertification Control Conversion.*

[6] Translator's note: The Dust Bowl developed from both human and natural forces. Overgrazing without crop rotation had depleted the soil; extensive and severe drought caused this powdery soil to become lighter. Winds whipping across the American and Canadian Prairie swept dark clouds of soil as far eastward as Chicago. A considerable volume of the soil blew all the way to the Atlantic Ocean.

[7] Translator's note: The Aral Sea, which is in the former Soviet Republic of Kazakhstan, has been shrinking since the 1960s, mostly as a result of poor water management in the region. Because the Aral Sea is salty, the sandstorms that blow from its shores are especially nasty: they cause much more severe respiratory irritation and crop damage in surrounding areas. Additionally, the salinity of the Sea has risen almost five-fold, killing most flora and fauna in its waters.

the relationship between different members of the ecosystem, we can achieve better results. Considering the diversity and stability of the ecosystem and the completeness of its function will help to bring about lasting change and the rebalancing of ecological processes. Indigenous plants must be emphasized in such planning. Starting with a critical mass of such plants, and allowing them to thrive, will revitalize the entire environment. Unfortunately, fencing off grassland accomplishes the reverse; grasses tend to dominate local plants. Worse, bringing in nonnative timber species can displace low-to-the-ground indigenous plants that cannot get enough light when surrounded by tall trees.

III. *Will Enclosures Improve the Grassland Situation?*

Closing off grasslands behind fences does not necessarily improve the situation of grasslands. Granted, short-term grazing bans can lead to an increase in ground cover and a corresponding decrease in wind erosion. However, more is not always better. Beyond some point, benefits can diminish. First, forcing an increase in ground cover through heavy-handed or ill-considered intervention will destabilize local ecosystems. Second, rapidly increasing ground cover and over-extending the volume of vegetation taxes soil nutrients, water supplies, and energy resources in a given system. All of these inputs are delicately balanced in grassland environments; pushing them out of balance is all too easy to do. Undernourished vegetation can rob indigenous and new plants of the soil minerals, water, and sunshine needed for survival. Thus, what seems like a good idea at first blush actually becomes a problem.

The Xiliguole Grassland Research Station, which is affiliated with the Institution of Botany of the Chinese Academy of Sciences, has closed off tracts of land to produce a perennial grass called *Leymus Chinensis* for about twenty years. During that time, total grass production and the number of grass species have declined. As of June 14, 2004, fresh grass per square meter equaled 113.14 grams, while withered grass amounted to 500 g. This compares unfavorably to land fenced off for only two years, which yielded 366 g of fresh grass against 234 g of unusable grass. Grasslands open to grazing produced 196.82 g of fresh grass, with 200 g consumed by livestock.

Note that after many years of protection, a substantial volume of grass had not really recovered. Grasses inside of enclosures remained yellow, even in mid-June. Worse, many plants that had been widely seen have disappeared. *Stipa grandis P. Smirn* has proliferated in its stead. This pattern mirrors what happened in Canada, where a plot of fenced-off grassland ultimately supported only one grass species. Decades of enclosure essentially caused the local ecosystem to collapse. This example serves as a warning that the road to Hell may be paved with good intentions.

Experience in Alxa, Inner Mongolia, confirms that the treatment can be worse than the disease. To roll back desertification, the locals planted timber trees in place of local sagebrush. This displaced herders and their livestock and produced a barrier between grazing rangeland and protected lands. Disturbingly, the protected areas have become relatively unproductive. Vermin—especially rats and various insects—have multiplied as well.

IV. *Can High-Yield Domestic Grasses Restore the Grasslands?*

Some persons believe that one *mu* of domestic grass can maintain one sheep. Some argue for wider use of domestic grass varieties as a way of modernizing stockbreeding. This will require a huge ramp-up in the amount of land used and the persons needed to maintain this land. While this may be possible and even desirable in some locations, for the vast drought-prone grassland such a model may prove vulnerable. Scarce wetlands may be overtaxed and water resources overtaxed. To maintain output, more land may be reclaimed, which in the long run will aggravate the situation. Barren areas at the edge of the Ke'erqin Desert have already been through this process. Banning grazing while opening up low-quality land can accelerate desertification and increase sandstorm frequency.

Differences of opinion on land use and grazing policy should come as no surprise. However, seeing the ban/enclosure model as a silver bullet for solving the desertification problem seems misguided in light of the evidence. Lifestyles and customs should not be sacrificed over this policy.

Too often, policies and people swing between extremes. Unfettered grazing and total bans represent the two extremes in this environment. Using one or the other will lead to a monochrome world of single grass species and one-size-fits-all lifestyles. Rather, policy should work toward outcomes that support natural systems. Diversity and harmony should be cornerstones of such policies.

This counts for natural as well as cultural outcomes. Too many "experts" fail to listen to the opinions of those belonging to minority ethnic groups. They blindly seek to refashion ethnic lifestyles and impose outside standards on local peoples. For instance, they pressure nomads to settle down on farms or in villages. This oversimplifies matters and leads to an extreme form of cultural assimilation.

Halting desertification, and even rolling it back, will not be done by going to extremes. Instead, different players in grassland ecosystems need to be harmonized and systemically integrated. Introducing newfangled, expert plans should be set aside in favor of approaches that hew to time-honored natural balances.

V. *Conclusion and Recommendations*

For thousands of years, the nomadic people on the Mongolian Plateau led a semi-collective and nomadic life. Grasslands and wild animals were communally held; individuals owned livestock. Divided into clans, tribes, and Gachas (administrative villages), they roved the grasslands. In spite of harsh conditions and scarce resources, these people have survived for millennia. Their personal endurance and cultural durability speaks volumes about their understanding of the land and how to live on it. Orders imposed from outside seem unlikely to improve local knowledge of how to weather droughts and live on what may seem like unproductive land.

Using methods brought in from the outside will probably not serve nomads well. For instance, domestic stockbreeding seems unlikely to provide a stable and substantial livelihood for locals. Forcing them into adopting these methods seems sure to bring about conflict.

Since before the Ming Dynasty, the Han Chinese have sought ways to quell the Mongolians. They built the Great Wall to keep out the Mongolian hordes. They limited grazing in certain areas. Han Chinese moved into the area in large numbers, introducing their farming techniques to the plateau. Rulers imported Tibetan Buddhism to pacify the locals. Interethnic marriages blunted tensions further. The occasional military threat kept Mongolians in line. In the face of this, the numbers of Mongolians and their livestock have dwindled.

Following the Revolution, new thinking came into play. Mongolians were encouraged to maintain their social structure and cultural practices. Locals reinstated economic rules regarding common and private property. Tibetan Buddhism, which had been introduced to Mongolia by the Qing Dynasty as a way of redirecting herdsmen and warriors to monastic life, was banned after the Revolution. Officials encouraged monks to return to secular life and become herdsmen, and accommodations were made to ease this transition.[8] This brought about a humane resolution to the problem of religion in Mongolia. In short, the grassland entered a new age.

During the Cultural Revolution this policy was reversed. Han Chinese, especially farmers, poured into the area. This pushed the grassland beyond its capacity. However, to maintain output of various agricultural goods, farmland expanded. Inner Mongolia began to look more and more like the rest of China.

The nomadic lifestyle began a rapid decline. Herders settled down, moving out of traditional yurts and into houses. The region became more and more agricultural and industrial. The government has increased investments in grassland restoration, mostly through efforts to increase grass production and plantings and tree planting campaigns. The aim is to forge a "new socialist pastureland." Officials put in place the Livestock and Grassland Double-Contract Responsibility System[9] to provide a policy underpinning to such efforts.

Through it all, the nomadic lifestyle—its values, its beliefs, and its organization—has declined. The culture has lost its dynamism and its resources. For instance, the number of Mongolian horses—a symbol of Mongolian military prowess and mobility on the plains—is declining. A distinct Mongolian culture has been largely overwhelmed by a common culture during this assimilation process.

At the same time, the local environment lost its stewards. Those who knew the subtle ecological balances and interconnections have lost the

---

[8] Translator's note: In ancient Mongolia there was freedom of religion. The most popular religion was Shamanism, a religion shared by many ethnic groups in the Chinese North and Mongolia. This changed during the Qing Dynasty, whose rulers imposed Tibetan Buddhism on the Mongolian people as a means of dominating them. The plan worked all too well: almost half of the male population entered monastic life. The resulting disruption impacted Mongolia economically by practically paralyzing the economy and disrupting the social system.

[9] Translator's note: Put in place in 1984, the LGDCRS stands on two pillars: Clementsian Succession Theory, the notion that inhabitants of a given ecosystem may in fact alter this ecosystem in ways that negatively impact the inhabitants (e.g., plants consume too much of the mineral base of soil), and property rights theory. The hope was that this system would encourage herders to keep livestock populations below local carrying capacities. However, the policy has yielded disappointing results.

knowledge, or at least the means to put their understanding to use. As a result, sandstorms have recently worsened and desertification-control campaigns have been called into question. Where once fierce warriors on horseback swept down form the north, now even more fearsome storms have taken their place.

Other ethnic minorities in China have suffered similar fates. The Owenk[10] Minority, a tribe in the extreme northeast, has suffered considerably. Although they enjoy the largest government subsidies of any ethnic minority, they have struggled to assimilate. When asked to disarm and move into suburban-style housing, they did so, but with disastrous results. Within a year, six young people died in the village in unusual ways. As the group declined, so did the herds of reindeer they had raised for centuries.

International efforts abroad have fared no better. For instance, the UN tried to assist various North African nomads by helping them drill wells and settle down. New villages, complete with high-yield crops and new livestock holdings, came into being. However, land around these villages eventually failed as the desert claimed the outskirts of town and its cropland. Variations on this theme have played out around the Sahara and as a result, the great desert has expanded.

In each case, the process started with ill-conceived human intervention. Outsiders, with little knowledge of the local land and ecological conditions, imposed their will on indigenous people who carry a deep, intimate connection to these fragile ecosystems. Perhaps the time has come for the experts to step aside and learn from the old masters. However, given the accumulated damage, can we actually put herdsmen and nomads back in charge of what was once their own land?

This depends crucially on how we plan for change and adjust the implementation of these plans. Solely relying on quantitative, economic, and material measures will likely lead to the same outcomes. However, if we look at conditions more holistically, qualitatively, and systemically, we may see a new path. This would lead planners and officials to consider cultural and ecological criteria in the planning and implementation of any change process. In short, the analysis of current conditions, desired outcomes, and the means for achieving them would be addressed comprehensively.

[10] Translator's note: Owenk is sometimes spelled Ewenke.

Such a comprehensive perspective would scrutinize a wider array of factors. All sorts of vegetation would be considered, especially indigenous varieties. Moreover, species diversity, not just total agricultural output, would move to the center of the analysis. Livestock, wild animals, and humans would be considered together, as well as the impact of invasive and opportunistic species. The interconnection of these elements would matter just as much as a raw census of their numbers.

Desertification in grasslands harms all of these individual elements. Plants struggle in shifting, unstable soil. Animals dependent on the plants, and animals that feed on them, face dwindling food sources. Culture, which shares a deep connection to the land, the animals, and the plants, gets pushed to its limits as resources grow scarcer. Successfully addressing desertification, therefore, requires that all of these elements be considered simultaneously and systemically. Addressing these problems in isolation, or the problems they produce, such as sandstorms, will not lead to lasting relief. In short, we cannot go to extremes, or address problems in isolation, and hope for success in restoring grasslands.

APPENDIX

NEWS CLIPS: THE FIERCE SANDSTORMS OF 2006

Fierce Sandstorm Causes a 32-hour Train Delay

At about 7 p.m. on April 4, the T-70 train from Urumchi to Beijing was swept up in a fearsome sandstorm. Blowing sand broke windows on the left side of the train. All seven hundred passengers and staff members fought against the onrushing sand for about twenty-four hours. The train finally arrived in Lanzhou at 9:41 p.m., thirty-two hours late.

News from Huhhot:[11] Sandstorms Increase Hospital Visits for Respiratory Distress

Frequent sandstorms and plummeting temperatures have pushed up the number of patients in the larger hospitals of Inner Mongolia. The number of persons seeking respiratory care has increased 50 percent over last year's numbers.

Severe Sandstorm Strikes Beijing

The worst sandstorm since the Revolution hit Beijing on April 17, depositing three hundred thousand tons of sand in one night on the city. By chance, The Sixth National Environmental Protection Convention opened that day. In an address delivered to attendees, Premier Wen Jiabao sternly warned that:

> We should all be aware of the need to address the sandstorm problem in Beijing. Right now, as we are meeting here on environmental matters, a massive sandstorm is lashing the city. There may be many reasons why this sandstorm is so severe, including climate change. Whatever the cause, this storm reminds us of the seriousness of China's environmental problems.

[11] Translator's note: Huhhot is the capital of Inner Mongolia, and is also spelled "Hohhot."

PART THREE

WATER

In *The China Environment Yearbook, Volume 2* we give particular attention to the issues of water security and access, quality, and rights trading.

Water is a strategic resource in China's sustainable development. It strongly determines the carrying capacity of various economic, social, and ecological systems.

Water safety issues take on many dimensions: economic, social, and ecological. Improving water quality and maintaining water quality standards stand out as key elements of water safety that cut across all of these dimensions. However, meeting standards has become increasingly difficult in the face of China's continued rapid economic and industrial growth and huge social shifts. Many rivers have dried up, lakes have been contaminated, shoreline waters sullied, and pollution nationwide continues to grow more acute, more varied, and more widespread. These factors have lead to ecological deterioration in many forms.

Climate change introduces new variables to the problem. Regional water cycles have been affected, which in turn impacts runoff and flow volume, especially of the rivers coming off the Tibetan Plateau and other highland areas in the West and South. These rivers are the lifeblood of Chinese agriculture, business, and culture. The longstanding imbalance between water supplies in the South and water needs in the North has grown sharper in the face of climatic disruption.

The lack of security in water supplies poses an immediate and growing threat to both economic and social development in some localities. This is not simply a matter of having too little water; water usage continues to be a problem that exacerbates scant supplies. For instance, in the East, water demand continues its sharp increase, which aggravates regional shortages, wrecks havoc on the environment, and increases the intensity, number, and types of water problems in China.

Reliable access to clean drinking water may be the most immediate concern regarding water in China. "Gaining and Maintaining Access to Safe Drinking Water," an article by Zhao Wengeng, elaborates on the prospects for improving access to clean water. Huo Daishan, the founder of a non-governmental organization named Huaihe River Guardians, made a series of investigations into the pollution and treatment of water in the Huaihe River Basin. His article, "Controlling Pollution in the Huaihe River Basin: Still a Long Way to Go," uses his firsthand experience to paint a shocking picture of water conditions in that area. Ma Jun's article, "Mapping Water Pollution in China: Informational Transparency at Work," directly links water quality conditions and the availability of public information about water quality. "Water Rights Trading in China," an article by Li Xi and Liu Mei, puts forward a novel approach to resolving water supply and distribution problems through the use of water rights trading.

CHAPTER THIRTEEN

GAINING AND MAINTAINING ACCESS TO SAFE DRINKING WATER

Zhao Wengeng

China suffers from one of the most acute, persistent water shortages in the world. The problem is not just a lack of water; in fact, only three countries have more fresh water than China. However, with more than 1.25 billion persons, and vast agricultural and industrial sectors, the country's per capita share is only one quarter of the world per capita figure.

Increasingly severe water pollution has dramatically intensified the water shortage problem. According to the China Environment Monitoring Center's June 2006 publication *Monthly Report on the Quality of Drinking Water Sources in 130 Key Cities*, sixteen cities failed to meet drinking water standards, and drinking water sources in seventy-four cities were substandard. Many rivers, lakes, and groundwater sources suffer from serious pollution. One hundred and forty thousand square kilometers of China's seas are polluted.

Unsafe drinking water has grown into a grave health hazard. According to a World Health Organization (WHO) survey, polluted wastewater causes 80 percent of all diseases and 50 percent of pediatric deaths. Over five million people in the world contract cholera, dysentery, and/or malaria annually. Over five million children die each year from water pollution and water-borne diseases.

Continuing to ignore these facts will imperil China's future access to safe drinking water, worsening an already difficult situation. Cities in the developed East as well as villages in the underdeveloped West face these problems. No one in China can deny the extent of the problem, especially in the wake of numerous water contamination accidents. In his comments on the widening water pollution problem, Mr. Shen Huaren, vice-chairman of the Standing Committee of the National People's Congress, said gloomily: "Now it seems water pollution is everywhere in China." He may well have been expressing the anxieties of the entire nation on this point.

Key words: safe drinking water, water crisis, water pollution

I. *The Challenge of Providing Safe Drinking Water in Rural Areas*

China has a rural population of nearly eight hundred million persons in more than two hundred million rural households. These people live in more than 2.5 million villages under more than six hundred thousand local administrations. Delivering safe drinking water to these individuals is clearly a daunting and complex task. Speaking at the International Water Conference, held in Beijing in mid-September 2006, government officials and water experts expressed their agreement on making rural water distribution a leading priority. At the same time, the Ministry of Construction held an International Symposium on Rural Sewage Treatment that recognized the linkage between sewage treatment and safe drinking water, especially for the rural population. Making this connection serves to build the "new socialist countryside,"[1] conservation goals, and development of an environmentally-friendly society.

According to a survey conducted by various experts and researchers, the Chinese face the following major problems regarding safe drinking water:

1. Currently, more than three hundred and twenty million people use unsafe water. In many places farmers drink water laced with high concentrations of fluorine, arsenic, and other contaminants. The North suffers from a chronic water shortage; the East has irregular access to high-quality water. The West suffers from both problems. The South, with abundant rain and river flow, faces fewer acute shortages.
2. Rural areas lack effective sewage treatment. According to the Ministry of Construction, 96 percent of villages lacked sewage treatments in 2005. On a related point, only 11 percent of villages offered refuse collection and processing services. Because of this, garbage is omnipresent in many villages, and a substantial portion of it finds its way into water supplies.
3. Funding problems hamper and undermine the construction and operation of water-delivery and sewage-treatment infrastructure. Small towns depend on government funds, property taxes, and other funds, while villages depend on special central-government fiscal allocations, collectively pooled funds, and individual farmers' money. In short, no reliable funding source exists that can provide sustained the financing needed to build large capital improvements. Resolving the funding problem may well prove difficult. Li Bing, director of the Village and Town Development Office of the Ministry of Construction, pointed out: "It will take a long time to find a complete solution to the problem. Money

[1] Translator's note: Realizing the goal of a "new socialist countryside" depends on several interlocking goals: poverty reduction; improved service delivery; social development; and cultural revitalization.

> alone cannot solve the problem." It seems likely that fiscal allocations will continued to be constrained because water pollution is directly linked to industry, and industry is seen as the engine of economic growth. Moves seen as tampering with growth will garner lukewarm support at best.

An example may help to illustrate these tensions and the impact of continued delay in tackling water pollution problems. A serious case of lead poisoning in Shuiyang Township, Huixian County, Gansu Province, vividly illustrates the extent of harm that water pollution can inflict.

Blood tests conducted on residents beginning in mid-August 2006 revealed widespread lead poisoning in the Township, especially among children. More than a thousand persons in two villages, Xinsi and Mouba, fell ill and went to Xijing Hospital in Xi'an City for treatment. Blood tests confirmed that these individuals suffered from lead poisoning. Children suffered the most; lead does more harm to children than adults. Children absorb 42 to 53 percent of the lead they ingest, about five times the adult rate. Making matters worse, they expel lead from their bodies at only about 30 percent of the adult rate.

Results of these tests exploded in the media, and government soon became involved. Officials from the Gansu Provincial Environmental Protection Bureau and a panel of five experts from the State Environmental Protection Administration (SEPA) went to Huixian County to conduct a detailed investigation. Their inquiry pointed toward a small lead-smelting operation near Xinsi Village, Huixian County Nonferrous Metal Company, Ltd., as the primary source of the pollution.

A more thoroughgoing assessment of the villagers' health ensued. The Vocational Health and Poisoning Control Institute of the China Disease Control Center found that of 153 blood samples, twenty-one had lead levels between 100 and 199 mg/L, thirty-nine between 200 and 249 mg/L, sixty-one between 250 and 449 mg/L, and one above 450 mg/L. In short, over 40 percent of the sample suffered from moderate to severe lead poisoning. All are children under fourteen; forty boys and twenty-two girls.[2]

[2] Translator's note: Note that the American Centers for Disease Control and Prevention (CDCs) consider a blood lead level (BLL) of 100 milligrams per liter as cause for concern. Clearly the BLLs recorded in this sample far exceed this concentration.

The government provided necessary treatment and compensation to all the victims. Local authorities closed down the smelting operation. Additionally, the government offered training courses to medical workers in the county, township, and villages on how to prevent and treat lead poisoning.

Pan Yue, vice minister of SEPA, pointed out that this example underscores the consequences of negligence on the part of local governments and the various environmental protection authorities in charge of oversight and monitoring. Given the scale of the damage, he recommended that those responsible must endure an administrative penalty. He also pointed out that such cases may be all too common.

However, many questions remain: Why was this factory built? Why was it placed so close to a major drinking-water source? Who gave permission for the factory to be built, and in such a hazardous location?

Answering these questions touches on many common, systemic problems related to pollution incidence. Hastily made decisions by local government officials in search of immediate economic returns are the root problem. While environmental and health impacts may be considered, the prevailing belief that "pollution can always be remedied afterwards," usually holds sway among such officials. Therefore, they readily sacrifice the environment, and their countrymen, for the sake of immediate, short-term economic growth. This lesson has been taught again and again, but it remains to be seen if it has been fully understood. Worse, those who feel the brunt of the lesson are not those who made the decisions; responsibility and impact are disconnected.

Building improved water-treatment and -distribution infrastructures in cities helps to secure safe drinking water for urbanites. However, country folk do not enjoy such benefits. Rural areas face multifaceted environmental problems, yet they lack the money necessary to address such difficulties. The extensive growth mode[3] works well for many urbanites in the East who enjoy the fruits of growth. Rural dwellers, however, do not enjoy these benefits, and they also shoulder many of the "external" costs.[4]

---

[3] Translator's note: The "extensive growth mode" is indicated by a focus on rapid economic growth, high energy consumption, and high investment. On the downside, these things come with large volumes of pollution, low efficiency production systems, low quality output, and tend to emphasize low value-added goods and services.

[4] Translator's note: By external costs, we mean costs outside of market mechanisms, especially price. These costs include pollution.

II. *Rehabilitation of Surface Water*

Fresh water makes up only 2.53 percent of the water on Earth. Glaciers at the poles contain roughly three-quarters of the total fresh water on the planet. For the most part, this water remains inaccessible. A considerable percentage of groundwater also remains out of reach. Therefore, we must use easily accessible freshwater—in rivers, lakes, and relatively shallow aquifers. Unfortunately, these water sources have become seriously polluted.

Chen Bangzhu, chairman of the Population, Resources and Environment Committee of the Chinese People's Political Consultative Conference, flatly stated at the end of 2005 that 70 percent of China's rivers and lakes were polluted and 75 percent of China's lakes show signs of eutrophication. Ninety percent of river water in urban areas is severely contaminated. Liu Changming, an academician affiliated with the Chinese Academy of Sciences and the head of the Water Research Institute of Beijing Normal University, pointed out that in data taken from 741 monitoring stations along the seven major rivers in China, none reported Level I or II water quality. Only 29.1 percent reported water quality at Level III, 30 percent at Levels IV and V, and a staggering 40.9 percent below Level V, or V-.[5] Water samples from two hundred lakes in eastern and southwestern China revealed that 80 percent endured eutrophication, which corroborates Chen's statement above. Given these data, one can clearly see that safe drinking water has become rarer, and therefore less accessible.

A. *Controlling the Huaihe River: Too Big a Challenge*

The Huaihe River may well be China's most complicated river system. It includes one hundred and twenty primary tributaries and more than four hundred and sixty secondary tributaries. One-hundred of them cross a provincial border, although the river mostly cuts through

[5] Translator's note: The Surface Water Quality Standard passed by SEPA in 1988 rates water in six categories. Level I is suitable for river source water in nature preserves. Level II is suitable for drinking water without treatment, swimming, Grade I Protected Areas, and rare and sensitive aquatic life. Level III is suitable for drinking water with some treatment, swimming, Grade II Protected Areas, and ordinary and unprotected aquatic life. Level IV is suitable for non-body contact recreational use and industrial use. Level V is suitable for agricultural and landscaping. Below Level V (Level V-) is completely toxic and unusable. Note that Grade I protected areas include river source areas and national-level nature preserves.

Henan, northern Anhui and Jiangsu, southern Shandong Provinces, and a corner of Hubei Province. Over one hundred and fifty million people depend on its waters.

Even though millions depend on it, and it cuts through a large swath of China's lands, it has become one of the most heavily contaminated rivers in only a few decades. Samples collected by the Huaihe River Water Resources Protection Bureau and Water Environment Monitoring Center from February 13–19, 2006, at forty-six cross-provincial branches of the river system, reveal the extent of the damage. Water quality at Level III was recorded at 17.4 percent of the stations, 10.9 percent at Level IV, another 10.9 percent at Level V, and 60.8 percent at Level V-. That is, 28.3 percent of the reporting stations in the basin recorded Level III or IV water quality, which are defined as "acceptable" in the State Council's Tenth Five-Year Plan for Water Pollution Control in the Huaihe River Basin.[6] An additional forty-three samples taken from the Huaihe River itself, the Grand Canal, the Yinghe River, the Wohe River, and the Shuhe River reveal even more alarming results. In the Huaihe's main watercourse, only one of thirteen sections met Level III water quality;[7] the other twelve fell into lower ratings. Ammonia and nitrogen ranked as the most common pollutants.[8]

Two decades ago, persons living along the Huaihe and its tributaries could draw water from the river for cooking and cleaning. Now, however, they must fetch water from miles away. Even that water must be treated with alum and stored for days in large jugs to bring it up to usable quality. It is tragic that so much water is so close and yet cannot be used.

A few years ago, many were shocked to find that people in Bengbu City, Anhui Province, had to drink bottled water even though the river ran right through the city. Today bottled water has become a common sight in many places along the Huaihe.

---

[6] Translator's note: Even though this looks like a relaxation of standards by moving Level IV into the "acceptable" category, in fact it is a way of measuring improvement. So much water in the basin is Level V and V- that the restoration plan places a high premium on any improvement above these dismal ratings.

[7] Translator's note: This one section is at Laobatou.

[8] Translator's note: This would suggest that a substantial portion of pollution comes from farm runoff.

In some sense, the Huaihe River has become the poster child for the damage that pollution can wreak on water supplies and the need to address these problems head-on.

The Huaihe's fortunes took a turn for the worse in 1994, when a major pollution accident occurred. In response to a persistent drought, all the sluice gates had been closed. Following a heavy storm in late July, the gates were reopened. Consequently, a huge volume of sewage moved downstream, polluting about 70 km of the Huaihe proper. This affected drinking water supplies and caused heavy economic losses.

In response, the State Environmental Protection Commission (SEPC)[9] under the State Council held a meeting along the Huaihe, stressing the need for more rigorous enforcement of environmental laws. Additionally, the SEPC advanced pollution targets to be met by 2001.

In March 1995, the SEPC, along with the Ministry of Water Resources and the State Planning Commission, held a joint meeting in Beijing to further discuss water pollution control plans for the Huaihe River Basin. Following this meeting, governments at all levels set detailed water pollution targets annually for the Huaihe River Basin.

At the sixth session of the third meeting of the SEPC, held on June 23, 1995, State Councilor Song Jian stated that the Chinese government considered the Huaihe River pollution problem a paramount concern. He put forward the idea that the polluting enterprises should be closed, have their operations suspended, or shift to other businesses as a way of substantially cutting pollution output. He recommended that laws and administrative measures serve as the primary tools for implementing such moves. Furthermore, he called on all parts of Chinese society to bring about change in the Huaihe River Basin.

The State Council circulated China's first river basin law, *the Provisional Regulations on the Prevention and Control of Water Pollution in the Huaihe River Basin*, in August 1995. The law, which took effect on January 1, 1998, barred industrial operations from directly discharging wastewater into the Huaihe River and its tributaries. It also placed caps on the amounts of various pollutants in industrial wastewater.

Later in 1995, SEPC reviewed the *Plan for the Prevention and Control of Water Pollution in the Huaihe River*. The *Plan* called for a 40 percent

[9] Translator's note: SEPC was dismantled in 1998, when the State Environmental Protection Agency (SEPA) took over many of its functions.

decrease in the volume of pollutants dumped into the river by 1997. It set an ambitious goal of restoring the river proper to usable water levels by the end of the century. To initiate this process, the State Council announced in September that all paper manufacturers in the region with a capacity of less than five thousand tons be closed by June 30, 1996. Other kinds of business faced the ax as well: officials shuttered 3,580 operations in the river basin, mostly small chemical and leather producers, in 1996.

On January 1, 1998, polluters that had failed to complete pollution control measures were closed down as part of "Operation Zero Hour."

That same day, the newly-constituted State Environmental Protection Administration, working with other State Council bodies, conducted an on-site check of the Huaihe's water quality. Soon thereafter, officials announced that Year 2000 targets had been met. Chemical oxygen demand (COD)[10] in the Huaihe's main watercourse achieved Level III, and COD levels in the major tributaries came in at Level IV.

Based on readings from thirty-three monitoring spots taken from September 10–15, 2002, 9.1 percent of the stations reported water quality at Level III, 12.1 percent at Level IV, 9.1 percent at Level V, and 69.7 percent below Level V. According to the targets set by the *Plan for the Prevention and Control of Water Pollution in the Huaihe River*, only 15.2 percent of the monitoring spots had acceptable water quality. The overall target was not met.[11] As of 2005, the Huaihe River proper averaged Level IV water quality, that is, lightly polluted, at least by the relatively lax standards applied for water in the Huaihe Basin. However, its tributaries did not fare so well. Sixteen out of twenty-four rated at Level V or V-. Water quality targets set by SEPA for 2005 had not been met.

---

[10] Translator's note: COD levels indirectly measure organic pollution in water.

[11] Translator's note: This requires some explanation. It is possible that the SEPA statement that 2000 goals had been met may have been overly optimistic. This may have been a hasty assessment, or it could have been the result of the manipulation of various industries and parties affecting river flow. As noted elsewhere in this article, firms and floodgate operators often hold back pollutants before and during inspections only to release them later. The other possibility is that pollution may have worsened in the Huaihe Basin the following year, but the degree of change calls this conclusion into question.

B. *The Costs of Economic Growth*

For over a decade, the central government and the governments in the Huaihe River Basin have spent vast sums on cleaning up the river and its tributaries. During the Ninth Five-Year Plan, the state launched two hundred pollution control projects and spent more than twelve billion RMB as part of this campaign. More than a thousand firms operating in the basin—such as the Lotus Flower Gourmet MSG Group and Fengyuan Group, two large but otherwise ordinary companies operating in the area—have spent tens of billions of RMB increasing sewage treatment capacity. More than one hundred cities, including Zhengzhou, Xuzhou, and Huainan, built or are building sewage treatment and refuse disposal and processing plants. Operation Zero Hour shut down numerous firms, cutting output by over twtenty billion RMB. Taken together, these four provinces—Jiangsu, Shandong, Anhui, and Henan—spent more than sixty billion RMB between 1995 and 2005 to control pollution and improve water quality. Simply put, this is not a trivial amount. And it is still increasing.

In spite of the massive spending on river-flow management infrastructure and pollution abatement, results have been disappointing. Effective pollution control and water management depend on more than just infrastructural improvements, pollution reduction campaigns, and legislation. It calls for a joint effort involving central and local governments working with companies operating in the area and people living in the river basin. It also requires that the elephant in the room finally be looked at: single-minded emphasis on hitting and/or exceeding GDP growth targets will not be compatible with achieving pollution control targets given the current level of technology, oversight expertise and manpower, and continued lax enforcement of environmental goals and laws. Pursuing growth at the expense of the environment will continue to generate serious problems and jeopardize economic development.

III. *Groundwater and Marine Pollution*

A. *Acute Groundwater Pollution and Poor Water Management*

At a talk given in July 2006, Dr. Wen Dongguang from the Department of Hydrogeology and Environmental Geology with the China Geological Survey announced that China would redouble its efforts in examining groundwater contamination. Wen stated that

> Groundwater in China is heavily polluted. This is clear. What is not clear is the degree and extent of the problem. Therefore, we need to conduct a complete survey of groundwater pollution. This will yield hard, quantifiable data on groundwater pollution and point the way toward solutions appropriate to the scale and types of pollution problems identified.

This survey will provide a long-overdue update to data currently on hand. These data come from a partial survey carried out in the 1970s, the results of which have offered only limited insight into the locations of the worst trouble spots, the sources of pollution, the content of pollution, and other key concerns. Without providing insight on these basic questions, attempts to address the groundwater pollution problem have lacked direction. As a result, work in this field has yet to deliver durable and significant improvements.

Contrasting this experience with the successes of the United States may provide needed insight in how to adjust Chinese efforts. The U.S. Geological Survey initiated the National Water-Quality Assessment Program (NAWQA) in the 1980s. NAWQA included a systematic investigation and evaluation of U.S. groundwater. By 2001, more than two-thirds of the aquifers in the U.S. had been surveyed. Among other data, information on organic and inorganic substances in groundwater had been measured. A second round of the program, begun in 2001, has extended the original results. Groundwater supplies threatened with serious contamination qualified for treatment under the Environmental Protection Agency's (EPA) Superfund program. The Superfund had a budget of about 1.5 billion USD in 2005.

Compared to these efforts, China's work in the area seems half-hearted and underfinanced, especially considering the scale of the damage and the potential harm to Chinese citizens. While we should not blindly imitate the United States, we could learn something about environmental protection from this effort.

Once fouled, groundwater is difficult and expensive to purify. It cannot be treated by conventional measures. To illustrate the point, a Japanese expert calculated the economic cost of treating polluted ground water at four hundred thousand heavily polluted sites in Japan. His findings show that a whopping ten trillion Japanese yen would be required to do the job.[12] Clearly, this goes beyond the bounds of

---

[12] Translator's note: Since the beginning of 1996 the yen has traded between 100 and 133-to-1 against the dollar. Therefore, in dollar terms this figure comes to between seventy-five and one hundred billion dollars.

economic feasibility. Therefore, we must prevent contamination to groundwater. Put another way: an ounce of prevention is better than a pound of cure.[13]

SEPA and the Ministry of Land Resources teamed up in 2005 to prepare a master plan to prevent and control groundwater pollution. The team is headed by Tsinghua University and includes experts from thirty notable institutions such as the China Geological Survey, the China Geological University, the Chinese Research Academy of Environmental Sciences, the China National Environmental Monitoring Center, Beijing Normal University, Jilin University, and others. The group expects to complete the work in 2007. The finished document will include detailed plans for the prevention and control of groundwater pollution in the next fifteen years. It will also include a groundwater survey; steps to be taken to protect groundwater sources, especially those used for drinking water; an outline of infrastructural improvements to protect groundwater sources; and policy recommendations, including laws, standards, and administrative processes, necessary to fulfill the overall aims of preventing groundwater pollution. When completed, it will serve as a basis for future work regarding the use of groundwater supplies, ongoing pollution control efforts, and overall groundwater protection and management systems and procedures in China.

On October 9, 2006, the 34th Congress of the International Association of Hydrogeologists (IAH) was held in Beijing. Over two hundred and forty specialists from fifty-six countries participated in the meeting, holding discussions on topics such as groundwater's role in sustainable development, groundwater safety and rehabilitation, rural poverty alleviation and groundwater use and protection, and exploration, evaluation, and use of groundwater. These discussions opened a new era for discussions of groundwater management in China.

B. *Oceans—the Final Frontier for Water*

Water is one of the basic elements needed for life. Life sprang from water on the Earth. Bodies of water team with life. Living things are largely made of water.

---

[13] Translator's note: Even with these costs, China's water situation has grown so dire that the Chinese Geological Survey began evaluating groundwater supplies in 2006 with an eye toward rehabilitating some of these sources.

Even though water is fundamentally important, human activity has sullied water supplies worldwide. This has threatened, and clearly harmed, a variety of life forms on the planet, including humans. Those life forms in need of fresh water have been especially hard-hit as pollution increases water scarcity.

To cope with this situation, various water treatment technologies have been developed, including desalinization of sea water. Unfortunately, the costs of many technologies, particularly desalinization, have put the technology out of reach for many countries and regions. However, as prices fall and technologies improve, they may provide reliable, ready access to clean water.

This begs the questions: Will desalinization work at an affordable price? When? To answer these questions, let us first look toward the sea, which contains more than 97 percent of the water on Earth.

The 2004 United Nations Environment Program (UNEP) report entitled *The State of the Marine Environment* identifies sewage discharge as an increasing threat to the world's oceans. The report noted that in many developing countries, 80–90 percent of sewage enters the sea untreated. This does not begin to describe the sheer volume and toxicity of pollutants contained in these discharges.[14]

On October 13, 2006, the Chinese Society of Oceanography held the third High-Level Forum on Marine Development Strategy in Xiamen. At the forum, Li Xiaoming, director general of the Department of Marine Environmental Protection of the State Oceanic Administration, said that around one hundred and forty thousand square kilometers of coastal waters in China had been polluted. Controlling and ameliorating this problem would require tremendous effort.

Beyond the sheer scale of the challenge, the complexity and variety of problems gives one pause. Coastal ecosystems, such as mangrove stands, have been damaged. Biodiversity has declined as species migrate out of their native habitats and overall populations drop. Various natural disasters, such as floods, have aggravated these and other problems. An already stressed and delicate ecosystem has been pushed to its limit.

---

[14] Translator's note: The report includes some bright spots as well. For instance, pollution caused by petroleum distillates and other chemicals, as well as persistent organic pollution, is on the wane.

C. *Concluding Thoughts*

Recall that the per capita share of water in China stands at only a quarter of the world average. This obligates all Chinese people and institutions to optimize water use by preventing water pollution and using water like the precious, life-giving substance it is.

Because of its geographical bounty—its large area and its access to the major rivers originating on or near the Tibetan Plateau—China can draw water from many sources. Plans to divert water as part of the South-to-North Water Diversion Project have been studied and may well be implemented to slake the thirst of China's parched North. Even with these advantages and technologies at our disposal, however, reducing water demand, using water carefully, and reusing and recycling water to recharge groundwater sources and for non-drinking purposes are but some of the ways that water supplies can be assured.

What we can and must do is to stop blaming others and shoulder the load individually and collectively. Liang Congjie, president of the Friends of Nature,[15] speaks the truth when he says that we must act with sincerity in tackling such problems. The words of Dr. Denis Hayes, founder of Earth Day, in 1970, second this notion: "As an individual, each of us can do little, but when we are united, we can save the Earth."

---

[15] Liang is the editor, along with Yang Dongping, of *The China Environment Yearbook (2005): Crisis and Breakthrough of China's Environment* (Leiden: Brill, 2007).

APPENDIX

News reports from Shuiyang Township, Huixian County, allow for the construction of a timeline that describes how the Lead Poisoning story unfolded.

On August 7, 2006, Guo Qin, a farmer from Xinsi Village, took his grandson to a hospital in Fengxian County for a blood test. This hospital was contracted to serve the workers of the lead smelting plant in the village and their families. After first refusing to conduct the requested blood tests, the doctors then completely reversed course and asked that the boy be hospitalized for "further observation," which aroused Guo's suspicion.

On August 12, Guo took his grandson to the People's Hospital in Huixian County, still seeking a blood test. Even though the test was normal, Guo regarded the result skeptically.

On August 18, Guo took his grandson to the Xijing Hospital for a blood test, which returned disturbing results: his grandson was suffering lead poisoning. His blood lead level (BLL) had reached 440 mg/L.

From August 20 to September 4, many of Guo's fellow farmers followed his lead by going to the Xijing Hospital for blood tests. As more results came back, the chilling evidence of widespread poisoning mounted. Among 879 people, 155 were suffering from moderate to severe lead poisoning; four had blood lead levels over 450 mg/L.

On August 22, the Huixian County government decided to suspend operations at the Huixian County Nonferrous Metal Company, Ltd., a small lead smelting plant operating near the village. Operators were asked to relocate the factory.

On August 26, more than eighty villagers staged a sit-down protest in front of the factory, since the factory had yet to stop production.

On August 27, more than two hundred villagers blocked the factory's gate when factory operators tried to move lead in powdered form and ingots from the site.

On August 28, the Huixian County Environmental Protection Bureau and the Power Bureau cut off the factory's electricity supply. In the afternoon, top officials in the county government met with three hundred villagers, finally deciding take decisive action.

On September 1, a panel of experts was sent by the provincial government to collect blood samples from the residents of Xinsi Village. The villagers refused to cooperate.

By September 4, 879 people had gone to hospitals in Xi'an for treatment, with more joining the process every day.

On September 8, SEPA and the Gansu Provincial Environmental Protection Bureau established a joint panel to conduct an investigation at the smelting plant and the surrounding area.

On September 9, the joint panel and the local authorities concerned dismantled the factory's furnace and shut down the factory completely.

By 4 p.m. on September 13, 258 people had been hospitalized, seventy-nine more than had been hospitalized on September 11.

On September 14, Pan Yue, vice minister of SEPA, commented on the nature of the problem (see quote in main text).

On September 14, the Huixian County government began to provide financial allowances to the victims.

On September 22, testing by the China Disease Control Center revealed that sixty-two children had moderate-to-severe lead poisoning. They were then hospitalized in the County Hospital.

Beginning on September 30, victims were treated according to the Proposal for the Treatment of Lead Poisoning in Huixian County. Fifty-nine poisoned children and twelve poisoned adults received treatment.

By October 8, 1,789 people had been tested. Of these, six hundred and twenty children had lead levels above 100 mg/L; forty-three adults recorded lead concentrations above 400 mg/L.

CHAPTER FOURTEEN

CONTROLLING POLLUTION IN THE HUAIHE RIVER[1] BASIN: STILL A LONG WAY TO GO

Huo Daishan

Huo Daishan, the author of the article below, provides a firsthand report of water and environmental conditions in the Huaihe River Basin. As one of the founders of Huaihe River Guardians (HRG), the first environmental NGO set up in the Huaihe River Basin, he knows the river and its challenges as well as anyone. HRG's "Project Hope: Saving the Huaihe River" aims to monitor pollution control efforts on the river and its tributaries. The project also includes a thorough investigation into the conditions of residents in "Cancer Villages." This investigation will form the basis for a "safe drinking water and medical care" campaign for people living in the Huaihe Basin. As repeatedly mentioned in the following article, the Shaying River, a major tributary of the Huaihe River, and also one of the worst polluted rivers in the whole basin, is a major contributor of pollution to the Huaihe River proper. Therefore, controlling pollution in the Shaying will directly impact water quality in the Huaihe.

Key words: pollution control, Huaihe River, Shaying River, water-borne diseases, public participation

I. *A Worsening Situation after a Decade of Pollution Control*

The Huaihe River is the third-largest river system and drainage area in China, running eastward and parallel to the Yellow River in the north and the Yangtze River in the south.[2]

Starting from its source in western Henan Province's Tongbai Mountains, the Huaihe River flows through Henan, Shangdong, Anhui, and Jiangsu Provinces. It spans a vast area: its length runs just over one thousand kilometers and its more than five hundred and eighty tributaries drain over two hundred and seventy thousand square kilometers (km^2). Over one hundred and eighty million people live in this river basin.

---

[1] Translator's note: The Huaihe River is also known as the Huai River in English.

[2] Translator's note: That is, it carries the third-largest water volume of any river system that is wholly within China.

The Huaihe is one of the areas where Chinese culture originated. Its abundant waters sustain a vast, fertile plain blessed with a favorable climate that launched a flourishing agricultural region many centuries ago. Wealth and stability allowed people to settle and develop a rich culture.

Some of the greatest of all Chinese thinkers hail from the region: Confucius, Mencius, Lao Tzu,[3] and others. Other legendary figures come from or settled in the region, including Da Yu,[4] the founder of the Xia Dynasty and the father of flood control techniques. Even the stuff of legend swirls through the landscape: Fu Xi,[5] one of the mythical Three Sovereigns of Chinese Culture, supposedly lived in this region.

The region is also known for its legendary floods. For various reasons, the Yellow River often flooded and its course changed. Over time, these floods washed an enormous volume of silt into the Huaihe River's estuary, blocking it and changing the Huaihe's course.[6] The Huaihe's waters eventually found their way to Hongze Lake and into the Yangtze River near Sanjiangying in Yangzhou City, Jiangsu Province. This split in the river's path and the overlapping flow of the three great rivers has come to be called "The Three Rivers in One Course."

The Huaihe became especially flood-prone because it emptied into the Yellow River rather than the ocean. A particularly severe flood swept through the basin shortly after the revolution in 1949. This prompted Chairman Mao Zedong to say, "It is imperative to control the Huaihe River!" China built many reservoirs and sluice gates on the river in order to realize this goal. By the late 1980s the Huaihe proper and many of its major tributaries had undergone improvements. River networks, man-made ditches, and irrigation systems brought bountiful harvests. Abundance brought great prosperity to the area.

According to the Huaihe Water Conservancy Committee under the Ministry of Water Resources, over five thousand seven hundred reservoirs, with a total capacity of 27.2 billion cubic meters, had been built in the area by the end of 2002. The basin included 5,427 floodgates; six hundred of these were medium-to-large-scale constructs.

[3] Translator's note: Lao Tzu is also known as Lao Tse or Laozi.

[4] Translator's note: Da Yu (or Great Yu) is also written Dayu.

[5] Translator's note: Fu Xi is also spelled Fuxi. He is credited with inventing and propagating writing, fishing, and trapping in China, and as such is one of the founders of Chinese culture.

[6] Before this, the Huaihe emptied into the ocean.

Over 50,000 km of reinforced dykes and levees lined waterways, and 2,164 km of artificial waterways and channels had been added to the natural river network. In addition, 60,480 electrical irrigation stations had been built. About 70 percent of the agricultural land in this area is irrigated by the Huaihe's waters.

Partially as a result of improved water capture and distribution, total grain yields increased from 12 billion kg in 1949 to 87.6 billion kg in 1999.[7] The area provides about one-fourth of the national grain output. Nearly one-fifth of the nation's cotton comes from the basin.

Even though water had become more plentiful, its quality suffered. Driven by industrial development in the basin, water quality rapidly degraded through the late 1980s and early 1990s.

Finally, the situation came to a head and the Chinese government took action. Authorities included the river in a major water control and rehabilitation project called "The Three Rivers and The Three Lakes Project."[8] The government took further steps, for instance by circulating *The Provisional Regulation on Water Pollution Control in the Huaihe River Basin* on August 15, 1995. This law clearly demonstrates the importance the government has attached to water quality in the basin: the law moved through legislative channels at near-record speed. Furthermore, as a law of the central government directed toward a regional problem, it shows the government's willingness to step in and exert its influence when the situation requires it.

Words met actions in the implementation of "Operation Zero Hour," which took place on January 1, 1998, and the "China 21st Century Environmental Protection Campaign."

After a decade of remediation, the question must be asked: How well have these efforts worked? That is, what is the state of water quality in the Huaihe River Basin in 2004?

Unfortunately, the ten-year anniversary of pollution control efforts in the Huaihe Basin got off on the wrong foot. A surprise rainstorm in mid-July 2004 produced problems in and along the river. As floodgates in Huaihe's upper reaches were opened, foul water from the Shaying and Wohe Rivers poured into the Huaihe's main flow. A cloud of effluent

[7] Translator's note: Obviously, other factors figured into improved output, such as increased use of fertilizers and other high-powered farming methods.

[8] Translator's note: The three rivers involved are the Huaihe, the Haihe, and the Liaohe; the three lakes are Taihu, Chaohu, and Dianchi Lakes.

more than 155 km long[9] carrying roughly three hundred and eighty million tons of pollutants stained the river. Pollutant concentrations increased seven-fold, forming a plume that killed everything in its path. Eventually the mess made its way into Lake Hongze, where it destroyed three hundred and twenty million tons of fish, shellfish, and other fauna and flora. Estimates of direct losses ran as high as three hundred and ten million RMB.

After all the work and investment, it only took one major mishap to set the river's rehabilitation back years. Needless to say, the media pounced on the story. Many outlets called the incident a "pollution rebound." Common folk simply saw waste and shortcomings in another grand plan.

The vast amounts of pollution released during this incident clearly indicate that the river and its tributaries had been fouled beforehand. In fact, of tens of thousands of firms operating along the Huaihe River and its tributaries, not a single company can demonstrate that it had met discharge targets. Those that had made "improvements" typically reported falsified figures to deceive authorities and the public.

Since the beginning of its pollution control efforts in 1994, authorities targeted and closed down five thousand enterprises. Unfortunately, the government targeted small operations, which produced a relatively modest amount of pollution. Large firms escaped unscathed, in spite of the fact that the types and volumes of pollutants they released could potentially "kill" the river.

In the ten years of remediation, development of sewage treatment and other pollution-control facilities lagged behind demand. During the Tenth Five-Year Plan (2000–2005), the government approved permits for 161 urban wastewater treatment projects. However, according to figures issued in 2004, seventy-eight had not even broken ground. Only fifty-five were under construction. A mere twenty-eight projects were completed, but these operated only for show when officials and supervisors visited them on inspection. When officials left, operations were suspended until the next visit or tour.

In addition to increasing industrial sewage output, household sewage output has grown with population and consumption. Compounding problems for the Huaihe, agricultural runoff has also increased as the

[9] Translator's note: The longest pollution plume recorded in the Huaihe Basin before this was 70 km long. It flowed through the river in 1994.

area has been called on to ramp up production in order to feed an expanding Chinese population and more ravenous global markets.

II. *The Difficulties of Controlling Pollution in the Huaihe River Basin*

The topic of pollution control in the Huaihe River Basin received considerable airing in public debates. While the public debated the finer points of what to do and when to do it, the work of carrying out programs had to proceed.

At a field meeting held in Bengfu City, Anhui Province, on October 23, 2004, vice premier Zeng Peiyan called on attendees to redouble pollution prevention and control efforts. Top officials from the four provinces along the Huaihe River responded by committing their governments to meet new pollution control goals within three years.

Fulfilling this "commitment" will be no mean feat. Because most of the pollution in the basin comes from industrial sources, pollution control measures mean extra costs for industrial operators, at least in the short run. Since few officials and administrators have the stomach or spine necessary to confront these operations, firms operate largely unchanged and unfettered. They still consume vast inputs; they still produce inefficiently; they still generate large volumes of waste and pollution. The emphasis remains firmly fixed on "the bottom line" rather than "the triple bottom line."[10] Even in those rare cases when the government applies pressure on firms, these operations often work out clever means of evasion or deception to continue working in a business-as-usual fashion.

Firms have a standard playbook for such trickery. A favorite is dumping discharge at night, when the probability of detection from locals or officials is low. Another is to discharge effluent on rainy days because open floodgates along various streams will more rapidly disperse pollutants. Still other firms simply put on a good show for inspectors and officials and immediately go back to dumping untreated effluent into river water after the inspectors and officials leave the company grounds. Some operators have installed devices that switch the flow in a given

[10] Translator's note: Whereas the traditional bottom line only measures financial costs and revenues, triple bottom line assessment brings together "profits, people, and planet" into a more comprehensive measure.

pipe from clean to dirty and back again. Other operations simply dump wastewater through longer sewage lines, placing discharge farther into the river's main current. Firms choose to engage in such mischief because doing so keeps costs low, or at least below the cost of taking effective remedial measures.

The following example illustrates the extent and impact of such deception in all its ugliness.

Ironically, on the day of the meeting in Bengfu City, vice premier Zeng inspected the nearby Yingsha River Floodgate in Fuyang City, Anhui Province. The water tested in this inspection was rated as Level II.[11] However, just north of the floodgate, the water had grown noticeably dark, reeking, and foam-covered. As it turned out, upstream discharge had been held back for twenty days in anticipation of the inspection. When word reached floodgate operators that the inspection had been completed, they immediately dumped what amounted to sewage into the river.[12] Water quality quickly plunged to Level V.

This low ranking seems more typical of the system. According to weekly water-quality tests performed by the Chinese Environmental Inspection Station, the Shaying River achieved Level IV water quality twice in both 2005 and 2006; every other test registered Level V or Level V-. Unsurprisingly, during the spring of 2005 pollution in the Shaying River showed no sign of abating. Columns of dark, stinking, foam-laden water have spread through the river, onto its shores and dykes, and even into cabin boats docked in its waters. Eventually, this tainted water will empty into the Huaihe. To combat this problem, the State Environmental Protection Administration (SEPA) adopted an

[11] The Surface Water Quality Standard passed by the State Environmental Protection Administration (SEPA) in 1988 rates water in six categories. Level I is suitable for river source water in nature preserves. Level II is suitable for drinking water without treatment, swimming, Grade I Protected Areas, and rare and sensitive aquatic life. Level III is suitable for drinking water with some treatment, swimming, Grade II Protected Areas, and ordinary and unprotected aquatic life. Level IV is suitable for non-body-contact recreational use and industrial use. Level V is suitable for agriculture and landscaping. Below Level V (Level V-) is completely toxic and unusable. Note that Grade I protected areas include river source areas and national-level nature preserves.

[12] Translator's note: One of the problems with maintaining water quality has to do with the fragile nature of water quality. As the saying goes: "If you take a spoonful of water and put it in a barrelful of sewage, you get sewage; but if you take a spoonful of sewage and put it in a barrelful of water, you still get sewage."

emergency pollution control plan for the Huaihe River in late April 2005.

At the beginning of 2006, the Bengfu agreement was adrift. Even though experience with the Huaihe Basin had grown, and the colossal Songhua River disaster had emphatically clarified the stakes of inaction, water quality reports made it obvious that conditions in the basin remained trapped in malaise. With more stations reporting Level V, the reality and gravity of the situation had become inescapable: things had not improved. In fact, they had stayed more or less the same, as noted in the consistently poor ratings of the Shaying River. With only one year left in the three-year push to improve water quality in the Huaihe Basin, we have no time to lose.

Unfortunately, updates from the Shaying brought even more bad news over the course of the year. A glimmer of hope came in June 2006, when the river's dissolved oxygen measured 0.59 mg/L, a record low. However, the real test would come in the fall during the flood season. When rains came and floodgates opened, waters turned even darker than usual; all fish exposed to these outflows died. Even for the Shaying, this kind of devastation had not occurred in a few years.

While the Shaying is only one river, it provides a shocking example of how bad water quality has become in China. Of course, water pollution is only one manifestation of a large complex of problems related to economic production, inefficiency, and relatively weak legal and enforcement institutions. Under these conditions, air pollution and soil contamination have proliferated as well. Mr. Zhou Shengxian, minister of SEPA, concluded that "The volume of major pollutants released into the environment has increased; we have yet to see evidence of significant and sustained declines in many cases."

III. *Avoiding Ecological Catastrophe in the Huaihe Basin*

Those living along the Huaihe and its tributaries have felt the brunt of this ecological devastation. The pain has been especially severe for those living near irrigation systems, since the wastewater discharged into rivers and streams will spread more quickly through the network of aqueducts and canals into farmland and ponds, contaminating soil, surface water, and groundwater. Groundwater contamination has a particularly deleterious effect on locals because they draw much of their water from ten-meter-deep wells. These and other wells less than fifty meters deep have been polluted in riverside villages. The water coming from shallow wells is typically reeking, murky, and silty. While boiling such water can kill microbial contamination, it will not remove sediments. Chemical contamination may be affected by boiling, but unfortunately, such chemicals can be carried away in water vapor which can prove harmful if inhaled.

The pollution problem in the Huaihe Basin became better known in the wake of a serious water contamination incident in mid-July 1994.[13] To gauge the extent and severity of the contamination, authorities collected three thousand liters of water from various places in the river basin and sent these to Shanghai for testing. Results identified 129 major pollutants in the water collected in the basin. Two particular samples registered ninety and ninety-five such contaminants, sixty-seven of which are carcinogens.[14]

Surrounded by water on three sides, Huangmengyin Village in Sheniu County, Henan Province, has suffered terribly because of water pollution. The Anhui Province Disease Control and Prevention Center collected water samples from two wells in the village in July 2004 to better determine the degree of risk faced by villagers. Results indicate that ten pollutants exceeded permitted limits, including heavy metals such as manganese. Nitrites exceeded limits by a large margin. In short, these results suggest that villagers are at high risk of developing cancer and other diseases.

A folk song from the Huaihe River region sums up the plight of such people in plainspoken, and acid, lyrics:

In the 1950s, we rinsed rice in the river,
In the 1960s, we used river water to wash clothes and irrigate land,
In the 1970s, the water quality turned bad,
In the 1980s, fish and shrimp had died,
In the 1990s, getting cancer is the fate of many.

People paid little attention to pollution before it threatened their lives. For many years, villagers did not know that they were contracting cancer

[13] Translator's note: This incident is described in greater detail in the previous chapter.

[14] Translator's note: The list of toxins, and their cancer-causing properties, have been established by the U.S. Environmental Protection Agency (EPA).

in record numbers. A mortality survey conducted in seven villages along the Shaying River from 1990 to 2004 shows that in each village dozens died of cancers. The most common types affect the digestive tract: esophageal, stomach, intestinal, and liver cancers were common and widespread. Some villages have "Cancer Streets" where every family along that street has at least one cancer victim. Huangmengyin Village provides one extreme example of this condition, where all four members of a particular family died of cancer. According to records from a local hospital, in 1972 only twelve persons had cancer in a population of 1.2 million. This roughly equals the world cancer incidence of one case in every 100,000 persons. Since then, cancer rates have skyrocketed in the region, rising as high as one in 100 persons in some areas.

In addition to digestive afflictions, conditions have affected the reproductive systems of both sexes. The number of sterile young couples and children with congenital deformities in riverside villages has increased over the past decade.

According to a 2004 survey, thirty-five children in Huangmengyin Village suffered from birth defects. Six from a nearby village had congenital heart disease. Many children in the area also suffered from various skin problems. Fortunately, these cleared up when these children left the area, but typically recurred when they returned.

Other afflictions had become commonplace. Fishermen and other persons working in local waters had sores on their hands and feet. The incidence of hypothyroid cancer, hypothyroid cysts, and breast cancer had risen to record highs. Various heart, brain, and circulatory afflictions became common among riverside folk, including young people. Though some of these diseases are more common among wealthy and elderly persons, the victims in these villages are typically middle-aged or young adults. Strikingly, they are also very poor; some get by on government relief.

IV. *Controlling the Pollution, Saving the River, Healing the People*

Water pollution has had a catastrophic impact on villagers living near waterways in the Huaihe Basin. Preventing and controlling pollution and delivering safe drinking water to these people has become imperative in improving ecological and health conditions in these villages.

Governments, non-governmental organizations (NGOs) and citizen groups have all worked together to realize this goal.

Huaihe River Guardians, an environmental non-governmental organization (ENGO), initiated Project Hope: Saving the Huaihe River Basin. This large-scale project seeks to

> Monitor pollution levels, identify sources of pollution, and supervise long-term pollution control efforts. As well, Project Hope will provide reliable access to safe drinking water to people living in polluted areas. Finally, we will facilitate medical examinations and expedite care for those living in cancer villages.

As part of this effort, HRG installed three hundred and fifty water purifiers in two villages between 2003 and 2005.

Other agencies have taken action. As early as 2000, Henan Province's government appropriated special funds to dig two 450-meter deep wells in two villages. New cancer cases in both villages declined from 1999 numbers. In the second year, new cases dropped steeply. By the third year, the rate of new cancer cases equaled national averages.

The central government pumped more funding into the area in 2005 with the creation of the Safe Drinking Water Project. This effort had a very simple but important remit: to provide safe drinking water for all persons in polluted areas. However, as early as April the project fell behind on meeting its goals: only 35 percent of planned projects had been completed and investments had only reached 42 percent of budgeted amounts. This triggered the implementation of emergency measures in order to get the project back on track.

Pan Yue, vice minister of the State Environmental Protection Administration, finally weighed in on the subject of water quality in the Huaihe River Basin. First, he flatly stated that water quality had not fundamentally improved. Second, and because of this, stated goals remained far out of reach. Finally, he went on to explain the causes of these shortcomings:

1. Provincial and local governments still shield many firms from feeling the force of environmental regulations. In some cases, officials turn a blind eye or simply do not look closely enough to find polluters; at the extreme officials conspire with polluting operations. Narrow, short-term gains can be funneled toward officials as direct monetary payments, or indirectly through status. Too many officials find such benefits too tempting to refuse.
2. Industrial restructuring has not moved forward quickly enough. Major polluting industries, such as food processing and paper,

textile, and chemical manufacturing, contribute over 70 percent to total wastewater discharge. The nation has yet to change its methods in these industries, move these industries away from major water sources, or move out of the low-margin/high-output segment in these industries.

3. The administrative offices charged with oversight and enforcement of environmental laws lack the manpower, administrative authority, skills, and financial resources necessary to fulfill their tasks.
4. Firms find compliance with environmental regulations more expensive than violation. This is a function of both the probability of detection, which remains low, and the severity of punishment, which is usually low as well.
5. Overlapping, vaguely defined, and unproductively-divided jurisdiction and authority among the various pollution control bodies leads to poor coordination in supervision, management, and enforcement of pollution problems.
6. The absence of robust prices has not allowed for rational production and allocation of recourses to take hold. For instance, without sound pricing, firms continue to allocate energy resources and various other inputs inefficiently. Moving toward markets should also increase demand for pollution control devices, which would improve the quality of operations among sewage and wastewater treatment plants.
7. Water resources are overexploited. Large numbers of dams and dykes have disrupted natural systems of water purification.

Controlling pollution in the Huaihe Basin will require successfully addressing all seven of these factors, both individually and systemically.

The entire basin must be regarded as an integrated ecological system. Successfully addressing problems in the basin will require environmental protection mechanisms that view the basin systemically. Exceptionalism and asymmetrically-applied laws will not deliver desirable and durable change.

Currently, however, the river basin is divided into administrative units and local governments that bear responsibility for only one part of the basin. One locality may be affected by upstream pollution sources, but unable to do anything about it. Other localities may know of pollution in their immediate area, but fail to control it because they will not feel the full brunt of its consequences.

These administrative differences can produce organizational gridlock. For instance, in 2005, SEPA found itself embroiled in a dispute with the Ministry of Water Resources regarding the total allowable volume of pollutants to be released into the Huaihe River. Even though these are central government bodies, they saw the issue at hand in fundamentally different ways. Moreover, they drew upon different data to form their conclusions. Systematizing and integrating data, improving interagency cooperation and coordination, and spelling out jurisdiction are only a start. Ultimately, a single authority, enforcing a single environmental policy, will have to take charge.

Local environmental agencies find it hard to stop new pollution from finding its way into the rivers and streams in the system because so many firms have nearly perfected the game of cat and mouse. The shortcomings of environmental agencies and the trickery of the firms they regulate must come to an end. The state's interests, the well being of the people, and the future of the environment are at stake; it is now time to get serious. Our future depends on it.

Achieving success will require vast resources, probably overwhelming governmental capacities. Therefore, broadening the public's role in this effort can speed up the process and bring new resources to these problems.

Public participation could consist of the foundation of a successful anti-pollution campaign. When people, plagued by pollution for years, are properly organized and act in an extensive, decisive, and timely way, they can accomplish a great deal.

The public can play a number of important roles. For instance, by using sheer force of numbers, they can provide round-the-clock monitoring of likely pollution discharge sources. They can collect discharge samples and have these tested by responsible and honest officials, ENGOs, and academics to make sure that firms meet official pollution targets.

Expert opinion has a role to play as well. For instance, many such experts call for the creation of "buffer zones," namely wetlands, between discharge outlets and rivers. This will help to filter water before it returns to a nearby river. This method can also be applied to household sewage.[15]

---

[15] Translator's note: Such systems are actually somewhat more complex. For instance, various plants, like daffodils, can be planted in such "wetlands" or marshes. These plants help to accelerate the purification process. They also help to protect soil quality and groundwater supplies.

ENGOs can lend their expertise in providing policy support and facilitating public participation. Such work has its challenges. For instance, many local governments take great pains to support and protect companies while ignoring NGOs, individual environmentalists, and even public input. Taking action in such conditions can jeopardize one's livelihood, home, and at the extreme, one's life. Only government can safeguard the public space for NGOs to be more involved in policies and for public participation concerning environmental matters. Even though government officials may find this a threat, this belief is misguided: engaging citizens and citizen groups in dialog can improve cooperation and the flow of information. Potentially explosive situations can be diffused before they reach a boiling point.

Controlling pollution in the Huaihe River Basin depends on a cooperative effort consisting of public participation with support from the government. ENGOs can play a pivotal role in building these new systems for tackling environmental problems. Success in this effort is imperative. The Huaihe River has played too important a role in China's economy, culture, and history to allow it to wallow in filth and pollution. Restoring the river could be a major success and steppingstone on the way toward building a conservation-minded, environmentally-conscious, and harmonious society.

CHAPTER FIFTEEN

WATER RIGHTS TRADING IN CHINA

Li Xi and Liu Mei

Water rights provide their holder with some control over how to use them and how to evaluate the underlying resources they represent. The rights give an incentive to hold, maintain, trade, and improve resources, depending on what decision provides the greatest benefit.

Typically, rights holders actually hold three kinds of rights: ownership, use, and profit. The first simply refers to having legal control over a given good or asset. The second refers to the rights of using that good or asset as the owner sees fit. The third refers to the rights of controlling and accruing benefits flowing from the use or exploitation of that asset or good, such as profit.

In China, ownership of water rights cannot be privately held since the state owns all water supplies. Therefore all other rights are conducted within the context of a lease or a license.[1] Therefore, the two remaining rights, use and enjoying the fruits of that use, will form the basis of our analysis here.

In order to enjoy some of the benefits of a market system, China has allowed rights holders to trade those rights. The system used, as well as the implications of this use, will form the main focus of this essay.[2] Water rights trading has led to great achievements and explorations in water use.

Key words: water resources, water rights trade/trading, water market

I. *Water Rights Trading in China*

Beginning in 1999 the Chinese government started to transform its water management policy from one based on administrative decree to a system grounded in market-based mechanisms. This process moved more to the center of water policy in 2000, when officials stated that they would pay greater attention to "water rights, prices and markets." The government hopes that this new approach will lead to an optimal use and allocation of water resources.

[1] Translator's note: This is similar to the way that broadcast spectrum is organized in the United States.

[2] Translator's note: "Water rights trading" has also come to be known as "water rights transference" and/or "water rights conversion." The latter name has been closely associated with water markets created in the Ningxia-to-Inner Mongolia exchange.

In 2006, China codified these changes into its regulatory framework through the twenty-seventh provision of the *Regulations on Management of Water Extraction Licenses and Water Resources Fees*, which stipulates that:

> Large organizations, public institutions and individuals who reduce water use by means of production restructuring, technological transformation, and water conservation are entitled to commercially trade the total water savings, provided they have the approval of relevant regulatory agencies. The Water Management Department under the State Council is responsible for implementing this system within a limited time period.

This regulation created the legal framework for water rights trading in China, but the practice had begun on a provisional basis in 2000.

A. *Water Rights Trading in Zhejiang Province*

The first water rights trade took place on November 24, 2000, between the neighboring cites of Dongyang and Yiwu in Jinhua District, Zhejiang Province. Situated in the Jinhua-Quzhou Basin in central Zhejiang, the two cities face completely different water supply situations.

Yiwu only has 1,130 cubic meters (m^3) of water per capita available to its residents, less than half the per capita water supply for the province in general. Total usable water in reservoirs amounted to only 150 million m^3. Before 1997, many citizens had to buy bottled water to meet drinking and cooking needs. Taken together these figures provide a clear picture of the inadequacy of local water supplies in the city. Insufficient supplies have become a drag on economic growth. However, the standard approaches for addressing these shortfalls—building additional reservoirs, expanding water supplies and capture, and redirecting more river and lake water to farms—will not address the root problems of local water shortages. Developing infrastructure within the area will help, but probably only in the short run. Lasting relief will only come by channeling water from a water-rich area to Yiwu.

One candidate stands out above the rest: Dongyang. Dongyang controls two reservoirs, Hengjin and Nanjiang. These provide the city with plentiful water. In fact, the town is able to release more than 30 million m^3 of surplus water into the Jinghua River annually.

By agreeing to trade water rights, development of the two cities became interconnected. The November 24 contract requires Yiwu's government to spend an estimated two hundred million RMB for access rights to Hengjin Reservoir's almost 50 million m^3 of water.

Note, though, that Dongyang retains ownership of the reservoir, and therefore retains responsibility for its management and maintenance. However, Yiwu must pay 0.1 RMB per m^3 of water to help defray such expenses. The total volume available, and therefore the annual expense, depends on the total annual water supply in Dongyang.

Yiwu will design, construct and pay for the water channel from Hengjin. However, Dongyang will take responsibility for the decision-making and construction of the water channel in its district, with funds from Yiwu's government.

The two hundred million RMB total was paid off over time. Each completed phase of the project earned a payment.

The project required three years to complete and now the Yiwu-Dongyan Water Channel carries water from Hengjin Reservoir to households throughout Yiwu, providing them with ongoing, reliable access to water.

This deal is tremendously significant because it moves infrastructure planning and execution out of the realm of administrative fiat and into the realm of negotiation and contracts.

B. *Water Rights Trading in Gansu Province*

In January 2002, the Ministry of Water Resources selected Zhangye City in Gansu Province in northwestern China as the first national-level water conservation test site. The trial period for this project lasted from 2002 to 2004.

Leading up to this, the State Council planned an emergency water management system for the Black River Basin. The goal was to develop and implement sound water distribution between the upper, middle, and lower reaches of the river system. This scheme required Zhangye, located in the middle reaches of the river, to discharge at least 225 million m^3 when water volume in the upriver reaches hit 1,580 million m^3. This scheme would also add 950 million m^3 of water storage across the river system.

After participating in the Black River and emergency conservation projects, Zhangye's leaders and citizens began to think about water resources more comprehensively. In particular, they wanted to achieve a balance between the river's natural demands and their desire for continued, rapid economic growth, even during water shortages. Through the negotiations leading up to the realization of this plan they gained especially detailed insights regarding water conservation, especially in setting up water rights contracts and trading systems. After a year, a no-nonsense system addressing these demands had been developed. They boiled down their insights to a simple rule-of-thumb: "Fix total volume, but adjust allocation among different departments." Implementing this aim required subsidiary supports. For instance, the city had to more carefully measure and control allocations. Of the total 2,600 million m^3 available, the city discharges 580 million into the nearby part of the Black River system and 950 million m^3 goes into the river proper to serve the basin's lower reaches. This leaves the city with 1,070 million m^3, which the city distributes locally according to rations for particular uses and departments.

The government extends the logic of this control mechanism to the area's water needs. Therefore, it sets quotas for different categories of use, primarily manufacturing, agricultural, and domestic uses, while setting aside enough to meet ecological needs. These quotas depend on activity and use in that sector. For instance, the allocation for farmers depends on such variables as the size of their fields and number of livestock. This allows for a daily quota to be set. Farmers then buy chits for a given volume of water. When they want water, they turn in these chits. This produces a paper trail and keeps the process transparent. Note, though, that if they can retain some tickets by conserving water, they can sell them on the market. This way, conservation is rewarded and while at the same time those who need additional water can acquire it legitimately.

Looked at systemically, this system rationalizes water distribution. Use moves toward efficiency; prices adjust in order to optimize value in the market. Moreover, by pricing water more rationally, it has a rationing effect. Farmers and other users realize the true value of water, and therefore, use it more intelligently. This prevents overuse and misuse; the former drives widespread shortages, the latter creates pollution. Underpriced water, or water with poorly-defined property rights, is prone to these kinds of overuse and misuse. This simple adjustment in water pricing helps to completely realign industrial, agricultural, and domestic operations. This, in turn, helps to restructure the urban and rural economies.

C. *Water Rights Trading In Ningxia and Inner Mongolia*

The latest case of water rights trading is found between Ningxia and Inner Mongolia in the Yellow River Basin.

Because the Yellow River Basin faces ongoing and sometimes severe water shortages, the need to develop new systems of conservation, allocation, and use takes on great importance in the region. The Yellow River Conservancy Commission (YRCC)[3] has taken up this cause, largely by helping to organize a water rights market along with input from the Water Conservancy Bureau. The basis of the system rests on trading water between the industrial and agricultural sectors. In April 2003 the YRCC began implementing this system by selecting five large industrial projects, including the Inner Mongolia Dalate Power Plant and the Ningxia Dam Power Plant, to participate in the system. Soon thereafter, it added two more industrial projects: the Inner Mongolia Yili Chemical Company and the Dafanpu Power Plant. These seven projects exist in two irrigational systems: the south bank of the Yellow River in Inner Mongolia and the Qingtong Gorge area. Laying the groundwork for the rights trading system went off without a hitch.

Inner Mongolia's government then selected the cities of Erdos and Baotou as the launch sites for the system. By the end of 2005, nine more projects had been added. These projects gained entry to the market through the auspices of the National Development and Reform Commission (NDRC). Investment in the system had climbed to eight hundred and forty million RMB. The system's capacity has swelled to 30 million m^3.

According to the *Blueprint for Water Rights Trading in Inner Mongolia*, the channel efficiency ratio of waterworks serving Erdos and the south bank of the Yellow River will save 130 million m^3 of water by 2010. This volume can serve some of the 1,300 companies waiting to join the water rights market.

Baotou City has also thrown its weight behind the development of a water market. It kicked off its efforts by integrating the Chengkouyangshui Irrigation Area into the market in stages beginning in 2004. As of this writing, two industrial projects have also been incorporated into the market. Note, though, that because an extensive irrigation system was already in place, these firms did not have to pay for additional improvements and water channels. According to the contract

[3] Translator's note: The Yellow River Conservancy Commission (YRCC) is an agency of the Ministry of Water Resources in charge of the Yellow River Basin and the inland river basins in Xinjiang, Qinghai, Gansu, and Inner Mongolia. The YRCC has responsibility for water administration in these basins. http://www.yrcc.gov.cn/eng/introduction/index.htm.

that they agreed to in cooperation with the administrative bureau of Chengkouyangshui's irrigation area, both projects would receive water supplies through the bureau's aqueduct system.

The Bureau of Water Rights Trading in the Inner Mongolia Autonomous Region has two other test projects on its docket. These have come to the stage when both parties are arranging a construction schedule, gathering funds, and allocating spending. The process has begun to take a familiar shape; various elements have been borrowed from previously-developed water rights markets in China. Systematizing these processes may allow for a set of standard operating procedures to be developed in creating these water rights trading systems.

Creating these markets has paid numerous dividends. Without drawing more water from the already beleaguered Yellow River, the trade has supplied water for various projects. Water shortages no longer impose a bottleneck on industrial development and economic growth. Farmers also benefit in the system: they bring in revenue by trading away water they save, which reduces their water use expenses.

The market-based system also helps to straighten out other kinds of water use and allocation in the area. Water is now used more efficiently, pushing China closer to its goal of effectively conserving resources, especially water. It also links the fate of industry to that of agriculture. This helps to maintain a more balanced approach to growth, and moves China away from a system where industrial growth has been achieved at the expense of the agriculture, the culture of rural areas, and the land itself.

The project also has greater organizational strength and flexibility than the administrative system. Funding the system does not require special, year-by-year allocations. This helps to ensure stability of the system. The 130 million m^3 traded in the market helps to bring in 1,000–2,000 million RMB to local coffers.

D. *Water Rights Trading in Hebei-Beijing*

The years-old dispute over water access between Hebei and Beijing has attracted considerable popular, governmental, and media interest. This row touches on some of the most crucial and fundamental aspects of water use and allocation: balancing regional and local interests; coordinating allocation between less-developed and highly-developed areas; and developing a workable system of regional development that considers natural demands and constraints. This deadlock finally opened up November 11, 2006, when the two governments signed *The Memorandum of Understanding between the Beijing Government and the Hebei Government on Strengthening Cooperation in Economic and Social Development* (the *Memorandum*). Negotiated over a six-month period, the *Memorandum* marks the end of water giveaways from Hebei to Beijing.

Even though the two sides signed the agreement on November 11, they actually began cooperating within the spirit of the agreement on October 13, 2007, when the Yunzhou Reservoir in Chicheng County in Hebei Province disgorged water to supply the Baihebao Reservoir in Beijing. This process was effectively repeated on October 15 when five reservoirs in northern Hebei, including the Shuigoukou, Xiangshuibao, and Huliu River Reservoirs, began to discharge water into the Guanting Reservoir in Beijing. Even though Hebei has supplied water to Beijing in this fashion for decades, this transfer occurred under an entirely new administrative and legal structure. The contract regulating these and future transfers called for more than water delivery; it required the two parties to cooperate in nine interrelated and overlapping ways. These include: transport infrastructure; water supply; environmental conservation; energy exploitation; tourism; and agriculture, among others. Furthermore, this should all be done under the rubric of social and economic "harmonious development."[4]

This new system replaces a system that privileged the interests of Beijing over those of Hebei.[5,6] In the absence of a clearly defined water rights system, Beijing, the seat of political power and wealth, drew water from Hebei. To some degree this was a matter of need: Beijing suffers chronic water shortages. Its per capita water allocation is only one-eighth of the national per capita, and only one-thirtieth of the global per capita. Being surrounded by Hebei, the city draws the vast percentage of its water, about 80 percent, from the province. Chengde and Zhangjiakou Cities supply most of this sum. By going into water

[4] Translator's note: "Harmonious development" stands out as more than a policy; it is an entire philosophy, cultural apparatus, social system, and economic style for growth, development, and living. More than just a feel-good notion of accommodation, harmony here is a vital force that interconnects systems and seeks to build strength through systemic balance and interconnection.

[5] Translator's note: Hebei surrounds Beijing, as well as another major metropolis, Tianjin. Both cities have drawn upon Hebei's resources, including water, to help power their growth and development.

[6] Translator's note: In addition to supplying Beijing's water needs, Hebei has provided 1,000 million m^3 of water to Tianjin annually since 1983. During some droughts Hebei even gave up its rights to nine hundred million cubic meters from Miyun Reservoir.

deficit themselves, these cities took on a variety of costs: money spent on environmental protection measures to protect the remaining water, including water pollution control; losses sustained in the agricultural and industrial sectors; and an overall slower economy.

The administrative priority system led to serious water misallocation in Hebei. For instance, during the worst droughts, Hebei had to purchase water from Shandong Province to meet its needs. During such droughts, shortages rose as high as 6,000 million m^3. To close the gap, the city pumped groundwater, which cannot be readily replenished and also causes sinkholes and other geological deformities. The extent of this damage can be seen in the newly-completed *Geographic Problems and the Blueprint of Mountains and Plains in Hebei Province*, which shows that 482 fissures have opened on the plains in Hebei. Seventy cities and counties in throughout the province have been affected.

II. *Comments on Water Rights Trading in China*

A. *Water Rights Markets and Efficient Resource Allocation*

Even though an effective, nascent water trading system has come into being in China, this market has yet to be fully developed. Because ownership rights are not fully recognized, markets will continue to be immature. However, this tool has proven its usefulness, and it should continue to be used. If anything, it should be used more widely and rigorously for a variety of reasons:

1. *Markets Improve Efficiency* Market-based trading systems will move water allocation closer to optimal efficiency. Negotiating market prices for water and the various goods and services served by water should balance benefits among the various market participants. This should diminish tensions and conflict over water allocation.

2. *Markets Generate Wealth and Greater Social Welfare* Trading water rights within a market-based system will increase social welfare. First of all, this system leads to more efficient use, which reduces waste. Secondly, by moving water from low-value to relatively high-value use, its use generates more value.

3. *Markets Can Support the Environment* Water rights trading leads to ecological improvements. As noted above, water is used more efficiently. As well, environmental NGOs and other water management agencies and organizations can participate in these markets. By doing so, they can purchase rights to hold water, assuring that ecological considerations are expressed through market mechanisms.

4. *The Civilizing Effects of Trade* Trade, because it depends on negotiation, levels the playing field among the various agents involved. Again, this helps to move China closer to the goal of realizing a harmonious society.

Relying on negotiations and prices to organize water transfers is a distinctly different process than achieving these transfers by administrative fiat. Although the old system has adjusted and improved, by discarding the completely-free transfer of water supplies between parties and by including symbolic payments and partial compensation, it still relies on coercion, for the most part. Under the old system the locale giving up its water simply could not refuse a government decree.

B. *Systemic and Technological Impediments*

Even though it has become more common and widespread, water rights trading remains in its infancy in China. It has gotten a foothold in the Yellow River Basin, largely because water allocation is such a huge issue there, but in most of the country it is little more than a good idea. This will likely continue to be the case until the basic issue of ownership is addressed. This and other impediments continue to undermine wider and more robust implementation of water trading systems:

1. *Incomplete Property Rights* Property rights remain vulnerable and under-defined in China. A fully-developed water trading system will only reach maturity when basic issues of ownership have been identified and resolved. Since the state owns water supplies, any usage of water or benefits flowing from its use remain in question. This murkiness has produced tensions between different water interests. For instance, until a rights-trading scheme came into play, Dongyang and Yiwu had come to loggerheads because the allocation system defined by the *Plan for Water Allocation in China's Yellow River Basin* failed to adequately address underlying rights.

2. *Administrative Disorganization and Jurisdictional Problems* Systemic and institutional overlap, disorganization, and poor coordination continue to hamper development of a water-trading market. Developing this market will be a long-term project, involving not only sellers and buyers, but market organizers and negotiators as well. Unfortunately, market organization and leadership in negotiation continues to be a sticking point. The management structure governing water use, allocation, distribution, pricing, infrastructure development, and so forth remains a morass of different agencies with different powers seeking different goals. Communication, cooperation, and coordination among different departments and levels of government remain spotty at best. Although the new water law has set up a system that combines drainage area management with regional administrative management, the old questions, such as decentralization of power, have not been settled.[7]

3. *Regulatory Content, Structure, and Integration* Building robust, efficient, durable, profitable, and effective markets relies on laws and regulations as much as it does on administrative and organizational structures. Without both, water trading markets will never fully develop. Currently, negotiated transfers have been narrowly defined and implemented on an ad hoc basis. For instance, when the YRCC approved the general planning report on Yellow River water rights trades in the Inner Mongolia Autonomous Region on April 1, 2005, it failed to ground this agreement in departmental regulations or legal documents. This oversight needs to be corrected in order for these negotiations to obtain full legal status, which will begin the process of integrating these methods into departmental standard operating procedures.

4. *Database Management* Negotiating water trades requires access to large amounts of reliable, high-quality data. For instance, rainfall, river flow rates, reservoir conditions, capacity of water transport infrastructure, amounts of water to be transferred, the means for transferring it, evaporation rates during transit, and prices for relevant variables, as well

---

[7] Translator's note: Coordinating the process of organizing markets and managing negotiations requires that present and future impacts be addressed through negotiations. Doing this effectively will improve efficient use of water in the future; failing to do so can sacrifice future use for narrowly-defined expediency in the present. In short, bringing together the right players with the right game plan leads to enlightened stewardship in the management of resources; failing to do so only uses the market mechanism to exploit precious resources in the name of profit.

as trends in variables that change over time, must be taken into account in considering rights trades. Based on these data, various calculations can be made to clarify current and future water needs, supplies, and prices in a given locale. Executing such calculations successfully requires that all parties have ready access to shared databases in order to make calculations consistent. Moreover, these data must be carefully entered and checked; a missed decimal point here or additional zero there can undermine an otherwise well-worked-out contract.

5. *Water Quality* Knowing and assessing water quality is of paramount importance in negotiating water rights transfers. For instance, if a given volume of water is transferred for a particular fee, but its quality does not meet a given standard and requires treatment, the recipient could find itself shortchanged in the deal. And because water quality changes for a variety of reasons, contingencies regarding water quality, and ongoing testing, must be built into contracts. Note that this testing requires equipment, as well as considerable manpower, expense, database management, and managerial acumen. These considerations must be folded into water trading contracts.

C. *Other Problems in Water Rights Trading*

The Ningxia-Inner Mongolia trade has provoked a specific set of problems beyond the sort enumerated above. These have attracted more attention and consideration from researchers, activists, policymakers, and administrative officials. Addressing these concerns is important in developing a more complete and standardized system for negotiating and implementing water-trading schemes.

1. *Lack of Context* Because water trades remain relatively rare, they seem to address a narrow set of immediate problems. This has run counter to developing a trading system that could be integrated into a broader body of policy that would move China toward better resource use and conservation. By failing in this regard, the benefits of a market-based system in preventing water waste have been forfeited.

2. *Contract Enforcement* Purchasers of water access have no guarantee that the amount of water stipulated in the trade agreement will actually be delivered. A number of variables can affect making good on this crucial part of any water trading contract: natural conditions like

rainfall, floods, and drought matter, as well as manmade impacts like contamination, failure of dams, dykes, or floodgates. Holding suppliers responsible for water delivery, even if their failure may have been the fault of a third party (e.g., an upstream entity releasing a contaminant into a water supply that affects the supplier's water quality). If the supplier fails to live up to its obligation, what recourse does the purchaser have?

3. *Infrastructural Fulfillment* Before water can be delivered, the system of delivery must be built. Because such projects are difficult, complicated, and costly, they can be delayed, they may suffer from ongoing problems once built, or they may fail altogether. Again, what legal recourse does the water recipient have in such cases? Note as well that many recipients pay for a lion's share of project costs. While the recipient certainly has greater motivation to complete the project, it is not clear if they will have the wherewithal to do so in a timely fashion. Providing low-interest loans or government-backed loans would go a long way toward providing such assurances.[8]

4. *Infrastructural Integration* In order to minimize direct project costs, many water delivery systems are designed using a "point-to-point" structure; that is, water is transferred as directly as possible from a certain point in the supplier locality to a specific location inside of the recipient locality. The problem with this approach is that it fails to integrate the new infrastructure into extant irrigation infrastructure.

5. *Rights Bundling and Unbundling* Thus far water rights trades have included multiple rights in a single contract. However, some deals may be negotiated more quickly, and may be more successfully managed, by breaking particular rights out of the comprehensive contract. For instance, developing a contract for sharing water extraction expenses and benefits could be a stand-alone deal, as could be a delivery and infrastructural improvement contract. This would allow for specialization in different kinds of water development and allocation work. It would also help to limit the number of government departments involved in

---

[8] Translator's note: Note, however that doing so would introduce some distortion to the market pricing mechanism.

a given deal. This innovation, in particular, may speed up the process of negotiating a given deal and putting it into operation.

III. *Policy Recommendations*

A. *Establishing a National Water Rights Trading System*

Because it has delivered real improvements, the water rights trading system is currently riding a wave of support. Now that a number of deals have been brokered and put into effect, this experience should be applied to develop a national trading system. This would require that a body of best practices be codified in laws and regulations to provide rock-solid legal, institutional, and administrative support for such a policy. This includes government playing a neutral role, dealing even-handedly with all parties, and maintaining transparency throughout negotiation, oversight, and enforcement phases of this work. Guidelines on the negotiating process need to be clarified, including language regarding contingencies that can complicate development of a project and delivery of water. Finally, the particulars of establishing jurisdiction, pricing, water quality, and sums to be traded need to be described and laid down. Parties to the process should always remember the primary aims of these efforts: improving water conservation and efficient use of water supplies; balancing regional interests; coordinating development across regions; and protecting the environment.

B. *Strengthening the Price Mechanism*

Water markets depend on rational pricing. Therefore, water prices need to be set by market forces. In this way, water reflects the price of finding water, capturing it, storing it, and delivering it. Engineering and infrastructure costs will be fully reflected in the cost of water. If markets are designed correctly and thoughtfully, even environmental costs will be reflected in the price of water. Note that in the North, where water remains scarce, it will have a higher price. With supplies lower, the good's price goes up. In addition, the infrastructure needed to deliver water, either through transfer projects or from wells, raises the cost of delivering water to users. In sum, price needs to reflect three considerations: natural resource value; management and development costs; and environmental cost, including treatment and rehabilitation expenses.

C. *Improving the Water Rights Trading System*

The water rights transferring system refers to the principles and methods of water rights trade. To improve the system, it should be incorporated into the development of a water-saving society. Regions engaged in water rights trades should make plans for building up conservation efforts in accordance with the mid-term and long-term goals of water conservation. This includes specifying goals, describing measures to achieve these goals, marshaling administrative and government institutions to work on these issues, and negotiating wisely among competing interests. This should help to foster the development of varied approaches to organizing and implementing water markets. The idea is to maximize total surplus, which will require careful planning and execution. Specific recommendations are:

1. *Increasing Contractual Flexibility* Trades can take different forms. For instance, contracts could be perpetual, long term, or short term. Contracts describing rental or leasing agreements could be adopted.[9] Transferring the value of rights into other tradable assets, such as stocks, or using other assets as payment for a water transfer, could be developed. Auctions could be used as a way of arriving at prices, rather than relying solely on negotiated prices for water supplies and different project elements.

2. *Clarifying Agreements and Contingencies* Even though every water trade will respond to unique needs, generally they will include particular agents who seek specific outcomes, and overall patterns will emerge. From these, a boilerplate contract should be developed. Experts in institutional economics, civil law, ecology, hydrology, engineering, public finance, and other related fields should be organized to accomplish this task.

---

[9] Translator's note: Although, in the absence of true ownership, the deals worked out in China amount to leases or licenses, as noted earlier.

Developing a more robust, comprehensive, and systematic water trading system can serve China's interests in numerous ways. Not only will the nation more successful manage one of its most necessary resources, it will also have at its fingertips flexible yet powerful market-based tools.

CHAPTER SIXTEEN

MAPPING WATER POLLUTION IN CHINA: INFORMATIONAL TRANSPARENCY AT WORK

Ma Jun

The Institute of Public & Environmental Affairs (IPE), a non-governmental organization (NGO) launched in May 2006, works on a wide-ranging remit. Their chief concerns revolve around water pollution. In particular, they focus on preventing and controlling water pollution, conducting research on environmental issues, and investigating transparency among environmental protection agencies and departments. This strong research and information gathering and processing focus of the IPE has led them to create a water pollution database and make it available to the public on their website. A central feature of the database is an interactive map that records and reflects data on water quality, wastewater discharge sources, and major polluters nationwide.

This searchable database can be accessed at www.ipe.org.cn. After entering the desired information in a search query, users can find pollution information about all thirty-one provinces and over three hundred cities throughout China. Various reports and summary statistics can be gleaned from these data in order to come up with rankings and lists, such as those of locales with the greatest volume of a given chemical in water sources, or of different contaminants in a particular area's water supply. Information on the kinds of chemicals in industrial, urban, village, and municipal wastewater can also be identified. The database also includes the Environmental Transparency Index (ETI), which summarizes the degree of transparency that various administrative departments working on water pollution issues apply in their work.

Key words: water pollution, Water Pollution Map of China, transparency

I. *The Role of Information in Controlling Water Pollution*

Total wastewater output continues its rapid increase in China. According to the *2005 Report on the State of Environment in China*, the total volume of discharged wastewater and sewage nationwide that year equaled 52.45 billion tons, Chemical Oxygen Demand (COD) reached 14.142 million tons, and ammonia nitrogen discharge climbed to 1.498 million tons. These are all sharp increases over 2004 figures. These discharges have tainted 70 percent of China's lakes and rivers. A disturbingly high 90 percent of urban groundwater is contaminated by organic and inorganic pollutants. This pollution emanates from fixed sources, such as factories, as well as multiple sources, such as households, farms, and even rainfall.

So-called "cancer villages"—small villages where cancer rates have grown to shocking heights—highlight the severity of pollution's impact on many citizens. Even though large numbers of people suffer terribly, research centers and government departments have failed to carefully and thoroughly examine the dangers of water pollution on human health. Moreover, what they do know about the linkage between pollution and health, as well as pollution sources and the degree of pollution in various bodies of water, has not been made available to the public at large. This lack of transparency is troubling: it leaves the public unprotected and undermines its confidence in the public sector.

Pollution, especially water pollution, highlights the tensions between economic growth and environmental protection. Moreover, dealing with this problem requires input and effort from numerous sources: government departments, research institutes, NGOs, citizens, and so on. Simply put, water pollution is one of the thorniest problems confronting China, and resolving it will require a formidable effort by many persons and organizations working in concert.

Enlarging the public's role in this work is crucial. Fortunately, China has been making progress in this regard. These efforts would be aided tremendously by having access to information regarding government operations in pollution prevention and control. By issuing the *Outline for Comprehensively Promoting Legislative Administration*, the State Council has gone a long way toward realizing this aim. Other laws complement this measure. The *Administrative License Law of the People's Republic of China* of 2003 grants citizens the ability "to request a hearing" on administrative matters. The fifth article of 2002's *Law of the People's Republic of China on Environment Impact Assessment* reads: "The state encourages relevant entities, experts and the general public to participate in the examination of environmental impact assessments in appropriate ways." Detailed regulations regarding the circulation of information used in environment-impact assessments (EIAs) are stated in the *Provisional Measures Regarding Public Participation in Environmental Impact Assessment*, a 2006 update of the aforementioned 2002 law. These rules and regulations enable the public to participate in environmental protection in a legal and orderly manner.

These rules and regulations have provided the basis of a legal foundation for gaining access to information and for holding government departments and officials accountable for their decisions and actions. Communications technologies, especially the Internet, further strengthen access to information and information sharing. Having access to such information is the lifeblood of almost any environmental protection campaign or organization. But even with this array of laws on the books, China still has a long way to go in broadening public participation in environmental matters.

Because information plays such a crucial role in understanding and tracking pollution, as well as in broadening and deepening the public role in addressing water pollution problems, we decided to focus our efforts on building and circulating a database on water pollution across China. Our aim is to aid public participation in water pollution prevention and control by arming the public with the best information publicly available. With this information, they can then work alongside environmental NGOs (ENGOs) and various government agencies to address water pollution and water management issues, to halt the continued erosion of water quality in China, and eventually, to restore water to a healthier state.

II. *Mapping Data: Making Information Easy To Read and Use*

To improve ease-of-use for these data, we used an interactive map as an interface. The China Water Pollution Map provides water quality and pollution information for all thirty-one provinces and over three hundred cities. The information mainly comes from bulletins, official government environmental-protection department reports, water departments, and oceanic administrations at all levels.

Data are accessible in an intuitive format. By clicking on, say, a particular province, a variety of information comes up: on water quality; waste and sewage discharge volumes; and the names of major polluters in that province. The map also details river systems and lakes for each province. A number of clickable buttons appear alongside the provincial map. Clicking these buttons allows users to explore water quality and pollution data on that province by year. Each informational category provides the following:

1. *A Water Quality Summary*

This contains information on the water quality of various bodies of water, including rivers, lakes, reservoirs, groundwater, coastal sea water, and drinking water by province.

2. *A Wastewater Content Summary*

The wastewater summary allows users to see what goes into local water systems and supplies. Data are organized by year. The database identifies numerous pollution sources. For instance, summary statistics on the volume and kinds of pollution coming from industrial and municipal sources can be found. Commonly-cited data can be found quickly, including: total wastewater output; concentrations of various pollutants in wastewater; percentages of various categories of wastewater that meet legal standards; and wastewater treatment rates.

3. *A List of Polluters*

By cutting the data more finely, the names of operations failing to meet legal standards regarding discharge can be found. This information can be categorized by province or city.

4. *Information by River Basin*

Understanding and addressing water quality standards requires that users organize data by river system. Since problems in one place impinge on water quality in another in the same river basin, we provide users with information on water supply, and quality bulletins and news on environmental protection efforts in a given river basin.

Maps from environmental protection departments, water supply and distribution departments, and oceanic administrations at all levels can also be retrieved from the database. These can be used to quickly see pollution conditions and supplement the categorical data listings described above.

III. *Mapping Public Actions*

By January 2007, more than three thousand two hundred polluters had been included in the database. Large corporations, including multinationals, have a major presence on the list, since larger operations produce more byproduct. However, most of the list is comprised of those companies that have been cited or otherwise identified by environmental protection departments for breaking environmental rules and regulations. A number of these violators have been required to correct these failings within a given time frame. At least this has been the case on paper. In reality, enforcement remains lax and spotty. Those that do obey the laws take on extra burdens, and those who do not simply evade detection, often with the tacit approval of official regulatory authorities. Because information on such malfeasance is hard to come by, and because information on meeting environmental standards remains tied up in bureaucracies, the *vox populi* has not coalesced into a coherent form. Therefore public and media responses have been muted regarding these abuses and shortcomings.

Making information on these and related issues available could empower the public, the media, and even the government. Legislators in the United States put this reasoning to the test in 1986, when they passed the Emergency Planning and Community Right-To-Know Act (EPCRA). Although this law did not directly address pollution levels, types, or impacts, its provisions made information from the manufacturers of potentially hazardous substances more widely available to the public.[1] Note that the law also opened up firms—and indirectly, regulators—to greater media scrutiny; the hope at the time was that this attention would lead to better behavior, or at least more vociferous criticism for large and/or repeated failures to meet environmental standards.

The act requests corporations to report the toxic and noxious substances they release into the environment each year to both the local government and the public. Based on this act, the United States adopted the first Toxics Release Inventory in 1988, listing approximately seven billion tons of toxic pollutants released from plants into the air, water, and soil. Media and environmental organizations have made excellent use of this information to shine bright lights on both the biggest polluters and the most seriously polluted areas. Advocates, organizers, and public agencies created websites to provide easy access to information on concentrations of pollutants in various locales. Interested parties could use this information to organize letter writing and fax protest campaigns and other actions, to press major polluters to change and government

[1] Translator's note: The U.S. Congress passed the EPCRA in 1986 under withering public pressure that had erupted after two similar, highly costly industrial accidents. The first one, which occurred at a Union Carbide pesticide plant in Bhopal, India, in 1984, claimed at least two thousand five hundred lives. In August 1985 a gas leak at another Union Carbide pesticide plant in West Virginia sent more than a hundred persons to the hospital.

agencies to take action against them.[2] The polluters soon felt enormous pressure from the public. To minimize the negative effects of these protests, they started taking action to reduce toxic outputs, hoping to be removed from the polluters list and to avoid public scrutiny.

The Inventory became an important legislative tool for the U.S. Congress. The list provided representatives information that helps inform the legislative process concerning environmental regulations. Regulatory agencies could now more effectively monitor and track the behavior of major polluters. This improved detection of polluting behavior, and therefore increased the likelihood of being sanctioned for such behavior. As the expected cost of bad behavior climbed, and as greater attention fell on those who flouted these laws, polluters began to feel the effects on their bottom lines and stock values. Once the formerly-external cost of pollution became internalized in such bread-and-butter terms, many corporations reformed themselves. This seems to have worked; the volume and noxiousness of toxins released in the environment in the United States has been in steady decline.

The State Environmental Protection Administration of China (SEPA) has requested that the list of polluters include more names and be more widely circulated. This has allowed the IPE to put more underperforming companies into its databases since 2004. We also hope that making this information public will prod local environmental protection departments to perform their monitoring and enforcement duties with far more rigor.[3] Our aim is that this information will serve to motivate the public to put more pressure on these firms. Under such pressure, companies will change their behavior, as was the case in the United States in the wake of the EPCRA's passage. Furthermore, we hope that consumers will use this information to make more informed purchasing choices; by supporting firms that practice sound stewardship, their good behavior is encouraged. At the same time, boycotting products made or sold by firms that underperform and/or engage in deceptive practices

[2] Translator's note: The EPCRA also established an emergency response system in case of toxic spills and emergencies.

[3] Translator's note: While the public role is emphasized in this paper, the government has a pivotal part to play in environmental protection efforts. Simply put, they need more power and resources to monitor potential pollution sources, and far more enforcement authority to sanction those who do not meet environmental standards. Note, though, that strengthening NGOs and the general public can actually aid various government agencies in this regard.

will hit violators where they will feel it most: in the profit margin. This signal should prompt them to reform their behavior.

To help publicize information in our database, we now work with a number of NGOs, such as the Beijing-based Green Earth Volunteers, Gansu-based Green Camel Bell, Henan-based Huaihe River Guardians, and Anhui-based Green Anhui to target enterprises and sewage plants that fail to meet pollution standards. With their help, we can pinpoint where the worst offenders operate and show this on our maps. Anyone with online access can then use our interactive electronic maps to inform nearby communities of the dangers they face, and at the same time help them to take action against such menaces.

IV. *Identifying the Pollution Hot Spots*

Users can use the IPE databases to rank provinces and cities according to various pollution criteria. When used in conjunction with head-to-head comparisons, users can get a clear idea of the degree of problems that exist as well as where they stand compared to other places in China. This should help the public, ENGOs, and the media to motivate change in addressing trouble spots.

Users can view rankings of all provinces or all cities. They can create profiles for specific provinces or cities. They can also create city rankings within a given province. Five types of rankings are available:

1. Total Wastewater and Sewage Discharge Rankings have four sub-categories: Total Wastewater and Sewage Discharge; Total COD Discharge; Total Ammonia Nitrogen Discharge; Per Capita Wastewater; and Sewage Discharge.
2. Industrial Wastewater Discharge Rankings have two sub-categories: Total Industrial Wastewater Discharge and Percentage of Industrial Wastewater Discharge Meeting Legal Standards.
3. Pollutant Discharge in Industrial Wastewater Rankings have ten sub-categories: COD; Ammonia Nitrogen; Volatile Hydroxybenzene; Oil and Petroleum Distillate Discharge; Cyanide; Mercury; Cadmium; Chromium; Arsenic; and Lead.
4. Urban and Township Sewage Discharge Rankings have two sub-categories: Total Urban and Township Sewage Discharge and Treatment Rate of Urban and Township Sewage.

5. Pollutant Discharge in Urban Sewage Rankings have two subcategories: COD and Ammonia Nitrogen.

The rankings show that some areas face particularly acute pollution problems. Hunan Province has the highest discharge of six different heavy metals. Guangxi Zhuang Autonomous Region must confront serious COD and ammonia nitrogen contamination. Shanxi Province leads all provinces in the volume of volatile hydroxybenzene in its discharge. Shaanxi Province tops the list in terms of oil and petroleum-based discharges. The causes of these concentrations of pollution will be investigated in the upcoming *Water Pollution Report of China*.

Public participation in environmental affairs depends on transparency. To support the public in this regard, we created the Environmental Transparency Index (ETI), which is based on our assessment of the degree of transparency in the operations of various administrative departments tasked with environmental protection work. We have included departments at various governmental levels: local, provincial, and central.

To construct the Index, we assessed three factors: openness of access to water quality information (thirty points); discharge information (thirty points); and information on polluters, including identifying polluters (forty points). We stratified our data by government level—city versus province—to make for more comparable groupings.

The first ETI shows that transparency remains largely unaddressed. However, with its ongoing updates, we hope that the Index will have a beneficial influence on highlighting the importance of transparency in administrative dealings. Our latest update was released in August 2006, and we believe that it will provide useful information to the departments under scrutiny, ENGOs, the media, and the public at large.

PART FOUR

FORESTS

This volume of *The China Environment Yearbook* also gives special attention to the protection of natural forests.

The World Resources Institute (WRI) released a *Report on World Forests* in December 2000 that analyzes worldwide forest conditions. The report stresses, "The world should consider not only the quantity, but also the quality of the forests," and that "in every developed or developing country, the problem is not the extinction of forests but the decline of old-growth forests."

This report hits on a crucial but often overlooked issue regarding the world's forests: virgin forests stand at the center of any meaningful discussion of biodiversity. This runs counter to the primary metrics used in forestry management, such as forest acreage, coverage rate, and timer volume; these measures cannot fully reflect ecological conditions. In short, we must move quality into the middle of discussion of quantity regarding forests. Furthermore, the notion of "greening" or increasing vegetation or tree cover is utterly superficial when such efforts rely solely on increases in single-species and/or manmade plantings. Immature forests can maintain timber acreage and coverage rates, but they may well mask deep-seated ecological disruption. Three articles in the following section wrestle with these issues head-on.

However, quantity has its place: as virgin and natural forests shrink and even disappear, we must use quantitative measures to raise the alarm about forest conditions.

Developing and implementing sustainable natural forest management stands at the core of raising forest quality. Shen Xiaohui elaborates on this point in detail in his article, "The Ecological Benefits of Improving the Quality of Forests." Feng Yongfeng's article, "Forest Rights 'Reform' and Natural Forest Protection," reveals that "reform" has resulted in the widespread clearing of natural forests and their replacement with new plantings that typically consist of nonnative species. "Chinese Wood Products Exports and the Illegal Timber Trade," written by Tamara Stark, Shi Pengxiang, and Cheng Yun, describes the timber trade and China's role in it, with a strong emphasis on the trade's distressing impacts on the world's old-growth forests. In each of these chapters, shocking and unsettling information is tempered by the increasingly helpful actions promoted by environmental NGOs, activists, and responsible academics.

CHAPTER SEVENTEEN

THE ECOLOGICAL BENEFITS OF IMPROVING THE QUALITY OF FORESTS

Shen Xiaohui

Many commentators claim that China's ecological system will substantially improve if its forest coverage rate rose from 20 to 30 percent. The latter figure is seen as a tipping point that, if met, would set China's woodlands on the path of a virtuous cycle of recovery. Some believe that building large, manmade, single-species forest is the most expedient and efficient means for achieving this goal. A forest cover target is even included in the Eleventh Five-Year Plan.

Including forestry concerns in basic planning makes sense. Forests play a central role in terrestrial ecosystems. They serve multiple and irreplaceable ecological, economic, and social benefits and functions.

After centuries of improper forest management, China's forests began to rebound in the second half of the twentieth century as total coverage climbed. However, while quantity climbed, quality grew shakier. Pestilence and increased desertification have threatened forests. Droughts, floods, and sandstorms have grown worse, largely because manmade, "mono-cropped" forests have replaced natural forests. Individual natural forests, especially old-growth areas, have grown smaller even as total forest area climbs. That is, the remaining natural forests are widely scattered, which seriously disrupts many of their ecological functions. In short, the forest environment fails to function up to its potential.[1]

Key words: forest quality, ecological health, sustainable development of forests

I. *Quality versus Quantity in China's Forests*

Chinese academics and government authorities have failed to adequately address the importance of quality in forest management. Without adequate definitions and guidelines, departments, policymakers, and even foresters face an only vaguely-outlined task. In the absence of qualitative guidelines, most players have fallen back on mere quantitative assessments. For instance, the *Sixth National Forest Resources Inventory* (hereafter referred to as the *Inventory*) evaluated Chinese forest quality

[1] Translator's note: The figures in this article mainly come from statistics contained in the National Forest Resources Inventory.

by leaning heavily on a census of the number of trees planted per hectare and forest capacity. The analysis also included gross figures on the growth of mature and partially-mature forests, as well as expansion in the coverage area of broadleaf, coniferous, and mixed broadleaf forests. Based on these *quantitative* measures, the *Inventory* concluded that forest *quality* had improved slightly.

A more robust definition of forest quality should consider the forests' inherent characteristics as well as their capacities. These capacities need to consider economic and social values, as well as the ecological contributions of a particular forest. Figure 17.1 helps to spell out some of these underlying factors, as well as how they relate to each other.

Developing such a Forest Quality Assessment Index (FQAI) could prove very valuable for China. Figure 17.1 clearly shows that a great many factors, and the relationships between them, matter in assessing forest quality. Gross figures, such as those in the aforementioned *Inventory*, exclude too much information to be regarded as a robust measure of forest conditions.

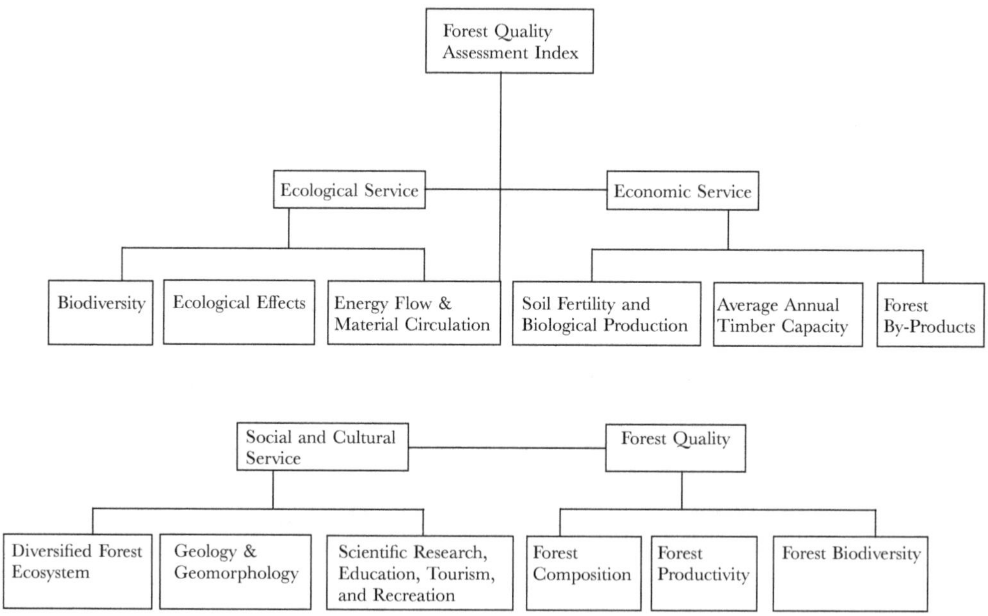

Figure 17.1: Forest Quality Assessment Index

A. *Modern Evaluation of Forest Quality in China*

Four to five thousand years ago, natural records indicate that forests covered 60 percent or more of China's current land area. Roughly six hundred million hectares[2] of these forests were virgin stands.[3] Only 2 percent of these forests, or about 5.5 million hectares,[4] still exist, according to Greenpeace's *Global Forest Map*. Most of these remaining forests are located in natural reserves or remote mountainous areas in southeastern Tibet, northwestern Yunnan, western Sichuan, the greater Xing'an Mountains in Inner Mongolia, and the northern edge of Aital Mountain in Xinjiang. Even so, these forests continue to shrink and become isolated pockets rather than forming coherent, contiguous expanses.

Chinese forests fall into various categories: virgin or old-growth forests; over-logged natural forests; natural secondary forests; and man-made forests. Each has distinct ecological and economic functions and characteristics.

Old-growth forest is typically the richest and healthiest of all forest types, at least in terms of ecological function and complexity. Numerous species, a near-optimal balance between kinds of species, and solid sheltering function set these forests apart from all others.

The remnants of old-growth forests that have been subjected to ongoing logging lose much of their ecological function. Although they still serve to stabilize land and provide shelter and ground cover, they have a less-subtly-balanced structure among species and among trees in different growth stages. With proper management they can recover many ecological characteristics of old-growth forests. However, heavy logging, especially clear cutting, does not allow these forests to recover any of their basic functions. At some point, this deterioration becomes nearly impossible to correct without heroic intervention. At the extreme, these areas become secondary forests or rocky alpine expanses. This has been the fate of the Yunnan-Guizhou Plateau. Under more favorable conditions, forests can recover to become secondary natural forests built around indigenous plant and animal species. If given time, they

[2] Translator's note: This equals six million square kilometers, or approximately 2,316,613 square miles or nine billion *mu*.

[3] Translator's note: Virgin forest is also called old-growth, primary, ancient forest, primeval forest, or frontier forest.

[4] Translator's note: This equals fifty-five thousand km^2, or approximately 21,236 mi^2 or eighty-two million *mu*.

can develop a healthy forest structure. However, these forests do not support the richness of biodiversity observed in old-growth forests, and lack some of the ecological function and stability of old, undisturbed woodlands.

There are a total of ninety-three million hectares[5] of natural secondary forests in China. Young-to-middle-age trees make up more than 70 percent of these forests. As the largest component of forest land in China, they play an important role in timber production, fuel supply, and maintaining local residents' livelihood. The function and great potential of natural secondary forests, however, have long been ignored. For this reason, substantial tracts of secondary forests have been cleared to make way for farmlands, manmade forests, or other purposes.

Protection and reforestation of natural secondary forests did not start until the implementation of the Natural Forest Protection Project, in 1998. By the time it was implemented, however, many natural secondary forests in China had been felled and replaced with fast-growing, high-yield manmade forests. While high-yield plantings produce immediate economic gains, they lack the ecological function of secondary growth forests. This oversight has begun to haunt Chinese forestry.

The tradition of planting manmade forests dates back five thousand years in China. In Huangling County, Shaanxi Province, an arborvitae stand in the legendary emperor Huangdi's Memorial Mausoleum was said to have been planted by the emperor, the ancestor of the Han Chinese. Since then, many manmade forests have been built. These have served a variety of goals: profit, shelter, fuel, and landscaping chief among them. Generally, as with the planting at Huangdi's burial site, these have been of small scale. However, as China has consumed more of its natural forests, it has developed more and larger manmade woodlands, especially since the Revolution. This practice has taken off under the aegis of the "farmland-to-forest" policy: new plantings have increased from a minimum of four million hectares[6] to 6,666,667 hectares annually.[7] With a total area of fifty-three million hectares,[8] China now leads all other countries in the total area devoted to man-

[5] Translator's note: This equals approximately 359,075 mi^2, or 1,395 million *mu*.

[6] Translator's note: This equals approximately sixty million *mu*, or about 15,444 mi^2.

[7] Translator's note: This equals approximately one hundred million *mu*, or about 25,740 mi^2.

[8] Translator's note: This equals approximately 204,634 mi^2, or seven hundred and ninety-five million *mu*.

made forests. These forests now play important environmental and economic roles. However, afforestation without sound planning, based on careful study of local natural conditions, can lead to a decline in forest quality.

B. *Major Problems of Forest Quality in China*

Beyond confounding quantitative with qualitative assessments, a variety of other challenges confront those working to improve forest quality.

1. *Young Trees, Immature Forests* Immature trees make up 77.4 percent of the total 53,333,334 hectares[9] of manmade forests in China, and 33 percent of total forest acreage. However, total timber capacity in these relatively young woodlands makes up only 8.28 percent of total forest capacity. Note that such forests reach only 28.4 cubic meters of lumber per hectare, or less than one-third of the lumber found in natural forests. About 6,666,667 hectares[10] of manmade forests have similarly low yields. Because trees are nonnative, or ecosystems have not reached a "climax" structure[11] in such places, trees often grow slowly, in spite of investments aimed at correcting this problem.

2. *Single-Species Plantings* Most manmade forests are comprised of single-species tree plantings. For instance, single-species coniferous forests make up two-thirds of the total; in some provinces this percentage reaches 95 percent. Afforestation efforts have been especially fond of fir, mason pine, larch, and Chinese pine species, which constitute more than 80 percent of the total area of plantings.[12]

Single-species plantings come with serious drawbacks: they tend not to support biodiversity; they are quite vulnerable to pestilence and disease; and they do not begin to mimic the better features of a fully-developed, climax ecosystem. Furthermore, over-reliance on single-species afforestation often overlooks considerations of whether

[9] Translator's note: This equals approximately 205,921 mi^2, or eight hundred million *mu*.

[10] Translator's note: This equals approximately 25,740 mi^2, or one hundred million *mu*.

[11] Translator's note: An example of a "climax" ecosystem is the "frontier" forest described below.

[12] Translator's note: Seventeen species are used in afforestation programs in China.

or not the species in question is appropriate for a particular locale. Finally, plantings lack systemic organization or coherence; no discernable overarching framework exists, other than putting as many trees in the ground as possible.

3. *Threats to Biodiversity* Artificial, highly simplified forest structure goes against and may well undermine biodiversity. China is one of the three most biodiverse countries in the world, so it has a special responsibility to protect this gift and legacy. All the more so since so many Chinese species face moderate to extreme stress: four to five thousand higher plants, ninety-four mammals, seventeen reptiles, and seven amphibians are on endangered species lists. Chinese species make up roughly 25 percent of the six hundred and forty species listed in the *International Convention on Trade of Endangered Species*. Fifteen to twenty percent of Chinese flora and fauna are under threat, which is much higher than the global rate of 10 to 15 percent. Habitat loss and degradation, hunting, and illegal land use all add to the pressure. A substantial part of the habitat loss and degradation, though, comes by way of single-species plantings.

4. *Vulnerability of Manmade Forests* Single-species plantings are more vulnerable and less durable than their naturally-developed counterparts. For instance, mason pine plantings in Chongqing dating back to the 1960s and 1970s have not yet grown tall or thick. Fire, wind, and as many as 359 diseases and insects have damaged Chongqing's woodlands. Nationally, 8,666,667 hectares[13] of forests have suffered from diseases and insect damage, killing four hundred million trees and destroying more than seventeen million m^3 of timber in recent years. These losses come to eighty-eight billion RMB, or about twelve billion USD. The Chinese have nicknamed this sort of destruction "the forest fire without smoke." Worse yet, these vulnerable forests continue to expand as plantings expand, adding 85,733 hectares[14] annually in the 1950s, to 365,2000 hectares[15] on average in the 1970s, to an astonishing nine hundred and forty-six million hectares[16] in 2004.

[13] Translator's note: This area equals around 130,000,000 *mu*, or about 33,462 mi^2.
[14] Translator's note: This area equals around 1,286,000 *mu*, or 331 mi^2.
[15] Translator's note: This area equals around 5,478,000 *mu*, or 1,410 mi^2.
[16] Translator's note: This area equals around 141,900,000 *mu*, or 36,525 mi^2.

5. *Poor Water and Soil Conservation Characteristics* Manmade forests, especially manmade coniferous forests, restrain the growth of lower level plants, which leaves soil relatively vulnerable to runoff. Furthermore, such forests do not capture water effectively, which undermines their ability in developing watersheds, and is one of the primary functions of a more naturally developed forest.

Two examples underscore the breadth and intensity of this problem. The first comes from Yunnan Province's Daliang mountainous region, where many pine trees sown in 1958 have grown no larger than 20 cm in diameter.[17] One possible explanation for continued slow growth is the lack of ground cover; the red, iron-rich soil is generally exposed to the air, making it vulnerable to wind and water erosion. A similar problem exists in Gucheng County, Hubei Province, which is famous for its manmade forests. However, water loss and soil erosion afflicts over 46 percent of the area. This outcome has occurred because trees were planted all over mountainsides. This overtaxed the soil, making it lighter and less cohesive, which accelerated erosion. Tons of soil were probably lost in the first year of the afforestation program alone. Research now shows that afforestation over an entire mountainside can cause four to six times more soil erosion than afforestation in only a single part. This negative impact on soil is common in any system of repetitive, single-species planting.

6. Single-species coniferous plantings do not hold water on or near the ground effectively. Pine needles do not provide as much ground cover on the forest floor. They break down more slowly than most kinds of leaves. Therefore, they do not enrich the soil and build humus as readily as other kinds of tree leaves. A rich environment for worms and other small organisms does not develop as readily, which deprives soil of enrichment and aeration. Dry pine needles also provide highly combustible material in forests; when combined with the resin-rich pine wood, they can fuel massively destructive conflagrations.

7. As a way of addressing these shortcomings head-on, some have proposed using genetically modified (GM) plants as the basis for forest plantings. Clearly, this is a risky course of action since the ecological

[17] Translator's note: These trees were not planted, but rather, germinated from seeds dropped from planes.

impacts of GM tree plantings have yet to be understood. In spite of this, over one million GM poplar trees have been planted in seven provinces. While genetic modification can help accelerate plant growth, the genetic material from these plantings can spread over great distances and contaminate natural forests, including old-growth forests.[18] Research into these impacts is underway in the United States, but only under very carefully-restricted conditions. For instance, plantings are limited to two acres. In the case of blossoming plants, flowering and access to animals attracted to flowers is carefully monitored and limited.

II. *Explaining the Low Quality of Chinese Forests*

A. *Outdated Forestry Management Theories and Tools*

Unfortunately, after 1949 Chinese experts modeled forestry management and policies on the Soviet model. Centralized planning focused on assessments of exploitation potential dominated thought and decision making.[19] For the most part, only two variables mattered: total timber and its net value. Planners and bureaucrats cast aside issues and benefits related to ecology and/or social impacts. Sustainability of planting and harvesting systems never entered into forestry policy-making or practice.

China began to develop an independent theory and policy practice in the 1960s, largely based on trial and error. In 1992, China changed directions again, adopting sustainable forestry techniques after the United Nations Conference on the Environment and Development published the most comprehensive and up-to-date information on forestry management. Sustainability has been the watchword in Chinese forestry theory ever since.

B. *Over-Reliance on Quantitative Assessments*

Economic growth in China relies on massive inputs, including natural resources. This has badly stressed Chinese forests.

---

[18] Translator's note: This process of contamination has been called "genetic pollution."

[19] Translator's note: The Chinese did consider alternatives, such as the German model of sustainable forest management, but failed to adopt them.

During the second half of the twentieth century, the nation's forestry system focused on timber production because of its ready availability and potential for use as an economic input. Over one hundred and thirty timber corporations were launched in this fifty-year period, with many enjoying some of the richest revenues of any large company operating in China at the time. In response, hundreds of thousands of people from all over China moved into forested regions in northeastern China, Inner Mongolia, and southwestern China. The demand for maintaining high output, and the lockstep coordination of the production system, suppressed selective logging and afforestation efforts. Unsurprisingly, this led to a rapid decline in forestland area and ecological quality.

Some of China's more extravagant political movements aggravated these problems. The Great Leap Forward,[20] the Dazhai model,[21] and the softening of regulations in the timber industry all negatively impacted China's natural forests.

From the 1970s to the 1990s, China placed a greater emphasis on developing forests, but still maintained an aggressive harvesting program. Between 1978 and 1988, total timber acreage and harvest potential decreased by one third. Put another way, consumption exceeded new growth by 1.7 billion m^3. By the late 1980s, timber companies faced serious financial problems. At the same time, the nation faced serious ecological problems as well, including flooding and desertification.

In spite of these overall trends, afforestation efforts had expanded. Three projects played particularly potent roles in expanding forests: developing shelter forests in the North; the national farmland-to-forest program; and a national afforestation movement. Implementation of these and related efforts can be broken into three categories: manmade planting areas; airplane sowing; and protected and/or fenced-off planting areas. These efforts added 4.2 million,[22] six hundred thousand,[23] and four million[24] hectares, respectively, of new forest.

[20] Translator's note: The Great Leap Forward was an economic program initiated in 1958 to accelerate China's industrial development, and to mechanize many productive systems, including agriculture and forestry.

[21] Translator's note: Dazhai is a village in Northern China that became the model of mid-1960s agricultural production after Chairman Mao identified it as the site of agricultural best practices worthy of imitation.

[22] Translator's note: This is approximately sixty-three million *mu*, or about 16,216 mi^2.

[23] Translator's note: This is approximately nine million *mu*, or 2,317 mi^2.

[24] Translator's note: This is approximately sixty million *mu*, or 15,444 mi^2.

Since the *Fourth Forest Resources Inventory*, total forest acreage and coverage has increased. With the boom in manmade forests, some ecological damage had been repaired. However, this "success" also meant that China's forests had become constructed affairs: manmade, secondary rather than primary, and generally of low ecological quality. In the long run, this condition will have powerful ecological, economic, and social ramifications.

The situation in Hainan Province clearly illustrates these trends. Early in the Qing Dynasty, primeval forests covered 80 percent of the province. Roughly 225 years later, forests only covered 26 percent of the province, for a total of 978,000 hectares.[25] Old-growth forests have dwindled to near-nothingness: surveys have shown that old growth has fallen from an average of 8.8 percent of the island's area during the 1990s to a mere 1.89 percent in 2006. That is, only 64,300 hectares[26] remain of the 331,000 hectares[27] available only ten years ago. In little more than fifty years, old-growth stands have contracted by 92.6 percent. Note that total forest coverage has increased in recent years, reaching 55.5 percent in 2005. However, crucial ecological functions, such as water capture and filtration, have been lost in the exchange.

C. *Forestry Management and Technology*

Ideally, forestry management considers ecological, economic, social, scientific, technological, legal, and cultural dimensions of forest growth, development, cultivation, and general oversight. As with many Chinese industries, commercial forestry has generally relied upon an "extensive" production model built on rapid growth, high short-term profits, large inputs, and inefficiency. Moving away from this model will require a top-to-bottom change in the industry, including some of the following steps:

1. *Improved Sprout Quality* The quality of sprouts largely determines the success of afforestation efforts and gross productivity of forests. Additionally, sprout quality plays a large role in shaping forests' long-term ecological and social functions. Developed countries use high quality

---

[25] Translator's note: This is approximately 14,670,000 *mu*, or 3,776 mi^2.
[26] Translator's note: This is approximately 964,500 *mu*, 248 mi^2.
[27] Translator's note: This is approximately 4,965,000 *mu*, or 1,278 mi^2.

sprouts in about 80 percent of their plantings; as of 2002, China lagged far behind, with only 28 percent.

Government regulation over sprout cultivation and marketing remains lax, allowing many inferior sprouts to enter markets. Fraud has also been a problem; sprouts claiming to be high quality are in fact nothing of the kind. Many indigenous species remained undersupplied by nurseries. This problem has arisen mostly as a result of market pressure: government authorities and commercial interests typically demand the fastest-growing species for plantings. Therefore hardwood and poplar trees now dominate China's northern timberlands; firs predominate in the south. Government and commercial interests tend to overlook species that provide the richest ecological function for a given environment, as well as those with long growing cycles.

With afforestation now adding more than 6,666,667 hectares[28] annually, neither the need nor demand for good sprouts has been higher. The variety and quality of sprouts planted now will undoubtedly determine the future of China's forests.

2. *Planting Manmade Forests in Lieu of Natural Forest Conservation* Indigenous tree species have many natural advantages over nonnative species planted in a particular environment. Unfortunately, planting nonnative species ignores these advantages and fails to make use of them. For instance, much of southern wastes covered by manmade forests could have been rehabilitated using less invasive methods. If this land can support manmade forests, it could have supported more natural forests, or at least plantings using indigenous species. However, because the nonnative plantings have taken hold, natural growth has been squeezed out and seems unlikely to return without Herculean efforts.

During the 1950s and 1960s, forestry managers introduced various eucalyptus species to China. Since then, these have become a favorite among forestry managers and commercial interests because of their economic productivity. More than two million hectares[29] have been planted; in Guangdong, Guanxhi, and Hainan, eucalyptus stands have been squeezed into almost every available area. They have even begun

[28] Translator's note: This is approximately one hundred million *mu*, or about 25,740 mi^2.

[29] Translator's note: This area equals approximately thirty million *mu*, or 7722 mi^2.

to move into Hunan, Jiangxi, Fujian, Zhejiang, Yunnan, and Sichuan. Mountains, coastal areas, and plains have been planted with them.

However, this is an invasive species; it competes for resources with indigenous tree varieties. Because it is so adaptable and hearty, it often dominates native species. This undermines biodiversity and generally undermines ecosystems. Restraints in importation and plantation of eucalyptus trees must be put in place.

Introducing eucalyptus species to the most fragile and specialized environments, such as those in Yunnan, has proven nearly catastrophic. To make space for eucalyptus plantings, southern subtropical evergreen broadleaf forests have been cut down at breakneck speed. An example of this kind of thoughtlessness can be seen in Cuiyun District in Simao, a city in Yunnan, which years ago had been called "a pearl in a green sea." However, this pearl was cast before swine. After dealing use rights to afforestation companies in the name of "property rights reform," these companies tore through 7,333 hectares[30] of natural forests in only four years. One company, Sinar Mas Group,[31] plans to plant as many as 1,733,333 hectares[32] of eucalyptus trees in the mountainous areas of southern Yunnan. Strictly speaking, these fast-growing forests function as cropland rather than forests. They exist to solve an economic problem; other considerations simply do not count.[33]

Eucalyptus plantings have grown increasingly controversial around the world. In New Zealand, and even in Australia, where the trees come from, environmental impact assessment must be executed before large eucalyptus plantings can go forward. Similarly, scientists from all over the world are probing the potential hazards of widespread eucalyptus plantings. New Zealand went so far as to cut down all eucalyptus for-

[30] Translator's note: This area equals approximately one hundred and ten thousand *mu*, or 28 mi^2.

[31] Translator's note: Sinar Mas' plans are detailed in "Memorandum Concerning the Asian Pulp and Paper Co. Ltd. (APP) Incident" written by Chang Li. This essay appears in *The China Environment Yearbook (2005): Crisis and Breakthrough of China's Environment*, Liang Congjie and Yang Dongping, eds., (Leiden: Brill, 2007).

[32] Translator's note: This area equals approximately twenty-six million *mu*, or 6,692 mi^2.

[33] Translator's note: So-called "forest rights reform" (FRR) involves one thousand five hundred counties and about 2.52 billion *mu* (around 168 million hectares or 648,651 mi^2) in those counties, being vulnerable to exploitation by commercial interests. Simply put, 57 percent of China's total forest coverage area should not be put on the auction block to be taken by the highest bidder or the savviest dealmaker. For more detail on FRR, see the next chapter in this volume.

ests. The World Bank has suspended loans and aid packages offering support to eucalyptus planting programs. In spite of this, eucalyptus planting runs rampant in China.

3. *Timber Production* The entire timber production and harvesting system needs to be thoroughly reconsidered. In particular, clear cutting needs to be constrained. Years ago, small clear cuts started in placates like the Changbai Mountain region. However, small clear cuts have grown into large areas. When this becomes the case, forest environments suffer deep disruption. Biodiversity takes a direct hit; large species generally have to flee clear-cut areas. Forests cannot readily regenerate, and soil erosion can result. The damage done to forest ecosystems impacts one of the most complex, mysterious, and necessary life support systems on the earth.

Selective harvesting has been implemented in Little Xing'an Mountain's Korean Pine stands, at least on paper. According to regulations, harvest areas should leave about 40 percent of a given area's canopy. However, canopy density has been reduced to 10 to –20 percent, partially as a result of trees being wind damaged or uprooted.[34] What was once knows as the "Home of the Korean Pine" has rapidly degraded into largely-denuded woodland.

Great Xing'an Mountain is covered by larch[35] forests. These forests have a remarkable capacity for regeneration. By conducting harvesting operations with care, forests here would prove durable. However, companies have not exercised such care: seedlings and young trees have been damaged by felled trees and tractors used to removed them. This damage leaves the forest vulnerable to soil erosion, which in turn undermines the renewal process. For these reasons, many mountains in this area have become barren.

4. *Over-Logging and Poaching* China uses a quota system to manage its limited forests. Quotas have been developed based on successful forestry protection practices that have been in place since 1949 and on best practices from aboard. The quota system and other forestry management tools have been codified into *The Forestry Law*, China's fundamental

[34] Translator's note: Korean pines have especially shallow roots, so they have less stability in strong winds. Harvesting thins wind protection, leaving remaining trees more vulnerable to wind damage.

[35] Translator's note: In particular, *Larix gmelinii Rupr.*

law on forest development and conservation, which came into effect in 1987. However, as with other environmental regulations in China, what appears on the books and what exists in reality diverge sharply.

According to *The Fifth Forest Resources Inventory*, over-logging destroys eighty-six million m^3 annually. As many as five-hundred thousand incidents of abuse are inflicted on forests annually. The most recent *Sixth Forest Resources Inventory*, which is the most up-to-date inventory, estimates over-logging at 75,542,100 m^3. Every year, the state invests nearly ten billion RMB, but only sixty-one million m^3 have been put in set-asides. Over-harvesting, tree poaching, and theft may produce the single most-damaging effects on China's forests.

5. *Lowering Standards to Make the Grade* The saying that there are "lies, damned lies, and statistics" seems especially relevant in assessing the quality of China's forests. Before publishing *The Fifth Forest Resources Inventory*, regulators changed technical definitions so that unforested wilderness, like bushland, is now counted as forest. This seemingly arcane change pushes 25,969,000 hectares[36] of general woodlands into the "forest" category. As a result, the minimum threshold for forest canopy in harvest areas fell from more than 30 percent to only 20 percent.

Changing standards does two things that should not be tolerated. First, changes play fast—and loose with forestry planning by putting the expedient ahead of the meaningful. Second, they make long-term comparisons very difficult. What looks like success in forestry management is really nothing more than an artifact of a decision made over what gets counted and how to count it.

III. *The Art and Science of Forestry*

Forests seem to evolve on a similar arc: they coalesce and develop, face stressors and degrade, and eventually recover and regenerate. Within this framework, we see a similar course for China's forests: we destroyed natural forests, we attempted to replace them with manmade forests, and finally, these will give way to the rise of natural forests. Along the way, China has moved from a doctrine of forestry based on total har-

[36] Translator's note: This equals approximately 100,267 mi^2 or about 389,535,000 *mu*.

vest volume and harvest potential to a method that honors the need to develop durable forests.

This slow, and incomplete, evolution has developed as a result of a deepening understanding of forest ecology. Through the 1970s, our understanding of forests remained superficial. However, in the 1980s, a more complex and integrated understanding of forest ecology began to emerge in developed countries. This had a deep impact on forest management, conservation policy, and environmental consciousness in general.

To improve the quality of the forests, I suggest that China make use of these insights to foster forestry policies and practices based on a deep understanding of ecology. Emphasis should be placed on redoubling conservation efforts, especially in protecting remaining old-growth forests; restoring the most troubled forest ecosystems; and developing forest ecosystems based on natural conditions, inasmuch as it is possible or reasonable.

A. *Protecting Natural Forests*

The World Resource Institute (WRI) defines a natural forest that sustains its native ecology as a "frontier" forest. These are the world's remaining, intact natural forest ecosystems. These forests are—on the whole—relatively undisturbed, and big enough to maintain their biodiversity and subtle variations in forest types. Unlike manmade forests that can be rapidly expanded and redeveloped, natural forests cannot be readily or willfully formed. For this reason, they require continued protection in order to keep them intact.

Eight thousand years ago, approximately eight billion hectares[37] of natural forests covered Earth's lands; the total forest cover now stands at 3.454 billion hectares, which includes manmade plantings and secondary forests. Forests contract twelve million hectares annually.

According to the WRI, China has lost 98 percent of its natural forests. What survives is fragmented and isolated, which reduces its ecological function and durability. Most of these forests are under protection, as parks or in forest protection programs, but a substantial percentage of

---

[37] Translator's note: This equals approximately 30,888,172 mi^2, or about 120 billion *mu*.

this land remains vulnerable to logging. Putting all natural forests into conservancy seems like an obvious, good move.

Improving forest quality begins with natural forest preservation. Clearly, this course of action will have substantial direct costs, but it averts an even costlier alternative. Simply put, this policy deserves greater support than the massive investments going toward developing manmade forests. Acting as if human beings can create a functional climax ecosystem is the height of folly and hubris; such ecosystems can only be protected and restored. This dovetails with the aims laid out by Premier Wen Jiabao, who points out that environmental and ecological protection should aim to protect, restore, and improve the ecosystem rather create one out of expediency.

B. *Restore Natural Secondary Forests*

Limiting or closing access to mountains and hillsides to facilitate the restoration of natural forests should be an important element of China's forestry policy. This builds on current policies aimed at encouraging forestation growth, but rather than attempting to achieve the best results through manmade plantings and other ham-handed efforts, this program would work with natural systems and designs. Doing so will help to realize numerous benefits: restoring natural beauty; attracting and supporting numerous animal and plant species; providing a suitable environment for microorganisms; and just as importantly, supporting conditions that carefully integrate and connect these elements into a subtle, powerful, durable, and organic whole. Human beings cannot achieve these outcomes alone; natural systems will more readily realize them.

Undertaking these projects will require a deeper form of observation and attentiveness, and a broadminded view of local conditions. Conditions like light, heat, slope direction, elevation, and size of trees all affect how these changes can be managed. The make-up of local flora and fauna also has to be considered. While expanding forests should remain a goal, it must be subordinated to the importance of fitting any intervention into local conditions.

Note that human intervention has a role in this process. At the very least, we should protect natural forests from logging. Beyond that, however, appropriate action can accelerate the recovery of secondary natural forests. For instance, by including edible plants in areas slated for restoration, intervention can help to sustain local incomes and diets and reduce the incentive for locals to harvest trees illegally.

Forests would be easier to revitalize if these areas faced less industrial disruption. Heavy machinery, trucks, and the roads that carry them have left their mark on forests in the last six decades, since China's lumber industry became more mechanized. However, relying less heavily on this technology would reduce the disturbance to the forest ecology, in particular its animal life. We now know that replacing heavy machines with lighter machines and manual and animal labor helps to achieve this sort of beneficial outcome. The Forestry Bureau in Heilongjiang Province has taken a leading role in reintroducing traditional lumbering methods in that region, with encouraging results.

Efforts like these have prodded forestry experts in China to advance the concept of "ecological lumbering." This approach attempts to minimize harm to the forest's ecological structure and function while still maintaining lumber operations. Thus while lumbering continues, it attempts to preserve biological diversity, variation in tree structure, and natural beauty. This method demands a systematic consideration of tree types, ages, and species, and attempts to maintain a healthy balance among these dimensions in order to maintain ecological stability even while harvesting trees. Even in logged areas, the basic environment seeks to simulate natural conditions: different kinds of trees coexist; fallen trees remain on the forest floor'[38] and a combination of coarse and fine debris line the forest floor, among other things. These elements provide an environment that nurtures a great many living things and provides the conditions for forests to renew themselves.

Additionally, larger natural forces, like wind and fire, must work on forests to ensure their continued health. These factors play crucial roles in regenerating forests, even as they may also cause great destruction. For instance, numerous forests, such as the natural forests in the Great Xing'an Mountains, require fire to renew themselves; forests in the Changbai Mountains rely on wind.

---

[38] Translator's note: Fallen trees turn out to be hugely important in numerous ways, such as holding water and moisture on the forest floor, blunting runoff in rainstorms, and providing a rich living environment for small animals and microorganisms.

In order to halt the process of degradation and to encourage the process of renewal, we need to widely implement ecological lumbering in China.

C. *Build more Manmade Forests that Simulate Natural Designs, Systems, and Processes*

Manmade forests cover 53,333,333 hectares in China;[39] plans to add 66,666,666 hectares[40] over the next ten years already exist. As noted above, however, quantity is only one variable to consider; for this planting program to truly succeed, it must abide by ecological principles and constraints.

Early afforestation efforts addressed a multidimensional and highly-acute crisis. Forest coverage had fallen to distressingly low levels. Resources, including financial supports, needed to remedy the problem remained highly constrained. Knowledge about best practices and ecological function remained scant. This privileged the "quantity over quality" outcome observed to date.

Even though mountains became greener, they failed to deliver lasting wealth to the denizens of the forests, including humans. In some cases, quality of life for those living near or in forests declined even as forest coverage increased. Numerous state-owned forestry centers borrowed money for afforestation efforts. This drove up public debts even as these artificial forests fell victim to pestilence, disease, desertification, fire, and other forms of destruction. In short, simple, mechanistic solutions cannot hope to solve complex, systemic problems. Continuing with "mono-cropped" forests simply wastes scarce resources. We must now adopt a richer system of afforestation.

To successfully fulfill this aim, we must change the means by which we measure progress and success in such programs. While total acreage, coverage rates, and survival ratios have their place, they only provide a shallow, gross look at forest conditions. Therefore, we must look deeper, by carefully examining healthy and optimal ecological function. This requires that forestry managers think systemically, balance various elements of forest function, and integrate different elements into a seamless

---

[39] Translator's note: This equals approximately 205,921 mi^2, or about eight hundred million *mu*.

[40] Translator's note: This equals approximately 257,401 mi^2, or about one billion *mu*.

whole. Rather than forcing an outcome on Nature, forest managers should listen more intently to Nature and follow its directives.

A few obvious measurements can serve as proxies for overall forest health. Measuring biodiversity may well top the list. Seeing multiple species in a shared environment indicates that many other systems of integration have succeeded. It also means that many smaller elements of forest function, such as the development of thriving microorganism populations, work effectively. Other broad measures, like soil stability and quality, can reveal a great deal about overall forest health and potential for recovery. Finally, weather conditions help to establish the forest's capacity for self-regeneration.

Leading international forestry research and development efforts place a high premium on naturalizing manmade forests. Germany has a three hundred-year history of managing manmade forests, and has continued to advance its acumen in this area. Note as well that Germany continues to operate a successful timber industry. In fact, the two efforts mutually reinforce each other.

The American Forestry Bureau adds to these basic insights. They recommend purposefully cultivating arbors, shrubs, and grasses for animal habitat. This attracts animals to forest areas and improves overall ecological function, provided that species remain native to the forests and animal populations remain at healthy levels. Attracting animals also depends on specific functions: eating, resting, breeding, and migrating, among other important actions. Given this, forestry managers must keep in mind that different kinds of forest land serve different functions.

Given the continued value of and demand for lumber, manmade forests for wood cultivation will continue to serve a useful purpose in China. Because of this, vast tracts of China's land will be devoted to meeting this demand. For instance, wood used for paper production will eventually cover five million hectares.[41] Unfortunately, this category of industrial forests is not suitable for use as quasi-natural forest. Still, we can take measures to enhance the diversity and stability of these lumber croplands. For instance, we can use more native species in such plantings. We should also move away from single-species plantings and combine conifer and broadleaf trees, and where possible and reasonable, interleave manmade and natural forests. However, these industrial manmade forests should be isolated from key biodiversity preserves.

---

[41] Translator's note: This is approximately 19,305 mi^2, or seventy-five million *mu*.

As the father of sustainable forestry, Hans Carl von Carlowitz, said in 1713:

> The greatest art, science, diligence and institution of these countries will rely on the manner in which such conservation and growing of wood is to be undertaken in order to have a continuing, stable and sustained utilization, as this is an indispensable cause without which the country in its essence cannot remain.

Nearly three hundred years later, his words still ring true. Von Carlowitz understood that forest quality is the key to durable forestry and durable forests. China now has the tools, the resources, and the need to put this thinking into practice. The health and wealth of the nation depends on doing so with sincerity and competence.

CHAPTER EIGHTEEN

FOREST RIGHTS "REFORM" AND NATURAL FOREST PROTECTION

Feng Yongfeng

In July 2003, the Central Committee of the Communist Party of China and the State Council formulated the *Decision on Promoting Forestry Development.* The *Decision* clearly outlined the process for accelerating reforms regarding collectively-owned forests. In response, provinces like Fujian, Jiangxi and Liaoning launched a series of reform measures aimed at refining this measure for local demands. This new system confirmed the primacy of farmers' roles in forest management, but within a broader process of devolving power away from centralized control. As forest resource managers, farmers would work to maintain and build their economic benefits, hopefully in a sustainable way. For the most part, these incentives have effectively raised farmers' incomes and wealth holdings. Ecological improvements have been realized as well. More fundamentally, farmers now hold titles to their land.

Pilot reform projects began in 2006, as government began deeding land. This ceded some power to farmers, and changed the balance of interest-group influence over forests. In the course of affecting so many people, disputes have appeared as well: some worry that the reform process will aggravate and accelerate damage to natural forests and even legitimize natural forest destruction.

Key words: forest rights reform, farmland-to-forest

I. *Introductory Thoughts on Forest Rights Reform*

In essence, forest rights reform (FRR) contracts farmers to manage what had been collectively-owned land. In other words, lands rights move from community ownership and control to individual ownership and control. This transfer is codified within a certificate of forest ownership. This transfer entitles the new owners to income generated from the land described in the deed. These two changes grant farmers vastly more autonomy over forestland management than under the former administrative system. This policy advances the Rural Household Responsibility System, which reallocates agricultural production materials and systems. According to Huang Jianxing, general director of the Fujian Forestry Bureau, this reform process extends new land policies to mountainous regions.[1]

Forest rights reform is actually a three-pronged system. First, claims divide mountainous lands among farmers through a lease contract. The second part introduces a "shareholders' claim" that allocates ownership of indivisible forestland to local households. Finally, a system of profit sharing is put in place to organize management, use, investment, and payments from land that remains under collective ownership.

This structure gives those who live in or near forests, and who rely on forestlands for their livelihood and way of life, good reasons to take a greater interest in protecting the forest's resources. By making these individuals and households responsible guardians of the forests, benefits and oversight become tightly linked.

This system has clear economic consequences and impacts. The new system builds on and extends the Household Contract Responsibility System (HCRS),[2] which has inspired millions of farmers' initiatives and produced remarkable economic achievements. While the HCRS formed an integral part of agricultural reform, FRR extends the process into the forest economy and the villages and towns in or near forests. As a poverty alleviation measure, the reform holds great promise, judging by the formidable track record of the HCRS.

FRR gets at some of the root problems that have stressed and damaged forests. Poverty sometimes drives locals to over-log wood. Poorly defined property rights leave resources unprotected and can give rise to classic "tragedy of the commons" outcomes. Together or separately, these forces can produce serious harm in woodlands.

---

[1] Wang Fanfan, Lin Yanxing and Li Qian, "China Will Adopt a New Round of 'Land Reform,'" *Xinhuanet*, May 14, 2006.

[2] Translator's note: The HCRS is a land use reform started by farmers and embraced and promoted by Deng Xiaoping during the 1980s. The system grants farmers a long-term license to farm on publicly-owned land. The license stipulates that farmers bear responsibility for the costs, risks, farming methods, and basic operation of farmland under contract. This system gives farmers an incentive to operate efficiently, since they retain surpluses for personal/household use and for sale on open markets. This has allowed many farmers to go beyond subsistence farming and commodity production to cash crop farming. Farmers pay a land tax, and provide a percentage of their harvests into collectively-held reserves. The HCRS has raised farm output and rural incomes. It has also released many persons from farm labor, allowing them to move into industrial production. The HCRS has become a cornerstone of overall economic and social change in China.

Zhang Chunshan, a volunteer environmentalist in Yunnan Province's Yulong County, illustrates these kinds of problems taking place near his hometown in recent years. He recalls that seven or eight years ago, a well-organized group used several trucks to poach hundreds of yews[3] on public land, and sold the wood in town. This has left a lasting impact: although the forest coverage rate shows no decline, the quality of the trees and forestland has declined. A large manmade planting has been made in these areas, replacing what had been abundant and diverse natural forest.

Locals did not regard the harvest of the yews as illegal since it took place on state-owned land. However, the *Forest Law* strictly forbids replacing natural forests with planted trees; such a replacement would not occur without first cutting down the native trees.

The location of these trees left them vulnerable. First, being on public land, no one felt any responsibility for them or for maintaining their well being. Second, because no one could effectively monitor their locations, marauders could cull trees without detection. This problem triggered secondary effects: generally poor land management set in and a devil-may-care attitude took hold. Forest fire incidence increased, but rather than provoking a response, it produced callous disinterest and fires were allowed to burn. At the extreme, some persons deliberately set fires in the forest to facilitate the growing and harvesting of mushrooms.

Zhang sees FRR as a means of addressing these sorts of problems, and for this reason he supports the reform process. In his eyes, the impact of the reforms is as clear as it is immediate: whereas previously a yew could be claimed on sight, harvested, and removed, that tree and the land around it now belong to a particular farmer. Zhang believes that these farmers will more effectively monitor activities on their land and take some pains to protect what has now become a valuable asset. This analysis basically echoes that of Chinese forestry authorities.

Jia Zhibang, general director of the State Forestry Administration, has noted that

> China has solved the problem of feeding its 1.3 billion people using 120 million hectares[4] of arable land. On the other hand, with over 285 million

[3] Translator's note: The yew is a large evergreen shrub (or small tree) that lives in upland areas. Its wood is hard and especially desirable in a variety of industrial uses, such as furniture manufacturing.

[4] Translator's note: This equals approximately 1.8 billion *mu*, or about 463,322 square miles.

> hectares[5] of forest land, we have yet to solve the problem of meeting wood demands; and we certainly haven't solved the problem of effective ecological protection of forests. The main reason we have fallen short is that forest reform has not been implemented. In the meantime, we face ongoing deficiencies in our forestry system and numerous impediments hamper the realization of productivity improvements.[6]

Even though the new forest rights management system builds on successfully-implemented reforms in agriculture, we must be mindful that agriculture and forestry operate in different ways and offer distinct economic and social benefits. So it follows as well that forestland and farmland require different methods of reform.

II. *Three Cases of Natural Forest Destruction*

For the purposes of this paper, forests are categorized two ways: natural and manmade. Each generates ecological and economic value. However, natural forests stand apart. They constitute possibly the most important land ecosystem on the Earth—no other system matches it in ecological benefits. Thus, these forests have value vastly exceeding pure monetary measures.

Simply implementing FRR can expose these vital forests to misuse and abuse. Utilitarian views, distilled as simple profit, can degrade and destroy these forests because manmade forest, which include more fast-growing and fruit-tree species, have a more immediate and clear economic impact. However, ecological "substitution" is nothing of the kind; single-species, nonnative plantings function as vastly inferior stand-ins for what Nature developed and nurtured for centuries. For instance, in northern Fujian Province, a site of FRR pilot implementation, economically valuable species such as firs, pines, bamboo, and fruit tress now dominate the mountainsides. The original natural forests have been almost entirely replaced. As a result, ecological conditions have deteriorated.

For better or worse, FRR has motivated farmers to become more engaged in forestry. The places leading the implementation of reforms

---

[5] Translator's note: This equals approximately 4.3 billion *mu*, or about 1,106,826 mi^2.

[6] Qin Jiaofeng, "The Forest Rights Reform Process Begins—An Interview with Jia Zhibang, General Director of the State Forestry Administration," *Biweekly Talk* 16 (2006).

have seen solid evidence of this. One such place is Fujian Province, where forest area doubled between 2004 and 2005, reaching about one hundred and thirty-three thousand hectares.[7] Afforestation in Jiangxi Province is no less impressive: from the winter of 2004 to the spring of 2005, 82 percent of about 222,333 hectares[8] of new plantings had been undertaken by individuals and private companies. Unfortunately, such impressive gains come at a steep price: damage to natural forests. Three cases below provide firm evidence of this problem.

A. *The Case of Fuzhu Village, Xinan Town, Xiapu County, Fujian Province*

Exercising their right to manage local forestland, Fuzhu Village residents and households, represented by local leader Lin Kaiji, entered into an agreement to lease about nine hundred and thirty hectares of secondary forestland to the Fangte Company. This mountainous land at the edge of town had been described as "barren hills" in the contract. However, the land contained ecologically valuable forests. The forests played a particularly important role in maintaining local water supplies. The vast gap in these perceptions—barren versus valuable—and the interests aligned with each point of view have sparked tensions. Fangte corporate representatives claim their lease contract clearly documents the legality of their claim and the methods of exploitation. Lin essentially backed the company's stance, saying,

> Our village has a large mountainous area and it would be a pity to waste it. Now we have put it to good use through this lease, which provides the village with a source of income. This income can be used for improving the well-being of the villagers.

Given the nature of the contract, the description of the land included therein, and the backing of village leaders, Fangte quickly went to work on clearing natural tree stands and replacing them with eucalyptus trees. These fast-growing trees can be harvested every six years to supply paper and particle board manufactures. Given the rapid growing cycle and increasing demand for paper and other goods derived from eucalyptus trees, such plantings help to fatten corporate bottom lines in short order. This strategy has paid off financially: the company

---

[7] Translator's note: This equals approximately two million *mu*, or about 515 mi^2.

[8] Translator's note: This equals approximately 3,335,000 *mu*, or about 858 mi^2.

has thrived in Fuzhu, and operates other tree plantations in the area, including one in Xiapu County.

The author traveled to Fuzhu in July 2006 to conduct field research on this case. Fast-growing eucalyptus covered the mountains by this time. The tallest trees were not yet three meters in height.

Villagers complained about these conditions and their ecological impact. One longtime resident related his feelings and observations as follows:

> The Xiapu Forestry Bureau played a trick on us and those who did not know better by claiming that those mountains did not have extant natural forests. Having grown up in the area, I had first-hand knowledge that broad-leaved, deciduous natural forests had thrived in the area for at least several decades. Decades-old American pines added to the mix. Large, mature trees, too big to put ones arms around, anchored the forest. When foraging inside the forest, the chill cut through my clothes and the canopy made the forest floor dark and foreboding. The dark, fertile soil that nurtured the indigenous trees now helps the eucalyptus trees grow very fast.

Traditionally, villagers drew their livelihoods from the nearby sea by fishing and fish farming. The small natural forests did not provide a direct economic boost, but rather ecological benefits. For the most part, the village and the mountainside existed in relatively benign coexistence. However, that changed when Fangte signed the land lease in 2004.[9] The forest had now become more of a cash cow in the deal: Fangte would pay one hundred and twenty yuan per hectare on a nine hundred and thirty hectare plot for a period of thirty years. Villagers believed that a key reason this deal went through is because Lin gained direct benefits from it. Some officials from the Xiapu Forestry Bureau shared this opinion.

B. *The Case of Jinghua Village, Hexi Town, Lanping County, Yunnan Province*

In August 2006 I traveled to Jinghua Village, Hexi Town, Lanping County, Yunnan Province, to investigate another case of natural forest destruction. Once again, official doublespeak claimed that "forest

[9] Translator's note: FRR has been rolling out in piecemeal form since 2004, but as stated earlier, the broader policy shift took place in 2006.

protection" was in place, but facts on the ground called this claim into question.

The Pumi ethnic enclave of Yushichang Natural Village lives in Jinghua Village. The Pumi have a deep tradition of protecting natural forests. Nevertheless, some villagers attributed their poverty to this tradition. As their attention turned toward making money, pilot forest-rights programs went into effect. Given their long and commendable track record in forest protection, Yushichang Village received an annual logging quota of 1,000 m^3 for 2004 and 2005. This brought in five hundred thousand yuan annually, which has swelled public coffers and pointed the way toward a more affluent future for the village.

Yang Jinhui, a local villager, described the complexity and difficulty of the situation:

> We take responsibility for protecting the forest and at the same time understand the privilege to make use of the forest. To do the latter well, roads need to be built for the transportation of logs, but this will accelerate the depletion of the natural forest. We have other problems as well: although the current policy allows only Yunnan pines to be cut, it is very likely that high-value trees like the yew and hemlock will be cut too. This happens because the forestry authorities have poor control at monitoring stations, and sometimes grant privileges to some people in exchange for a favor to be called in later or for an interest in logging operations. Measures used to protect the forests have been perverted as well: people sell wood in their cooking fuel and construction quotas for cash. Trees belonging to rare species have been culled. A key problem is that natural forests are left unprotected by being categorized as "barren land."

Yunnan Province's natural forests have tremendous ecological value. But even here, forest reform has been used as a cover for cutting down native trees and replacing them with fast-growing "cash crop" trees. Government officials in similarly-endowed provinces have contracted tens of thousands of *mu* of natural forests to paper mills as part of the "forest-pulp-paper integration" system.[10] These piecemeal projects add up: at least half of the land used in recent afforestation efforts had been thriving natural forest before being cleared and replanted with nonnative, fast-growing species.

[10] Translator's note: Ten thousand *mu* equals about 667 hectares, or about 2.57 mi^2.

C. *The Case of Nanping City, Fujian Province*

Similar conditions obtained in Nanping City, Fujian Province, another FRR pilot site. As usual, the "reform" operated under a confusing babble of bureaucratic doublespeak: "bidding on mountains" amounted to buying the rights to natural forest timber and ongoing rights to plant high-yield tree varieties in their place; "forest upgrading" or "low-yielding forest improvement" advocated the same, but within a rhetorical framework of increasing revenues and profit. The height of verbal legerdemain was expressed in the catchphrase "greening barren hills," which simply overlooked the existence of long-standing natural forests; such thinking regards mountainous land as a blank slate to be rewritten with the stylus of industrial forestry's methods. The particular market structure created in each case will likely speed the diffusion of manmade forests in lieu of what had been collectively and/or publicly owned and managed. The effectiveness of these policies in creating new plantings and new revenues has clearly been demonstrated. Without a doubt, the bidding system used has increased local incomes, especially those of officials in charge of negotiations. However, the ecological prudence of such a system remains in doubt.

This tension can be seen in what advocates of FRR see as two of their more significant success stories: Fujian and Jiangxi Provinces. At 62 percent of its land area, Fujian has the highest forest coverage rate in the country; Jiangxi is second highest with, 60.95 percent. Clearly the reform process has helped create tree cover. However, natural forests have suffered terribly. What little remains falls into three categories: areas set aside in nature preserves; areas serving symbolic or health purposes, such as *fengshui* forests;[11] and areas that have religious significance (typically these receive only marginal management and care). Lin Peng, a professor from Xiamen University and a member of the Chinese Academy of Engineering, pointed out that the natural broad-leaved forests in Fujian account for less than 10 percent of the total forest coverage, or less than 6 percent of the total land area. Single-species coniferous plantings, eucalyptus stands, and fruit tree groves make up the bulk of the tree population. He warned, "Although Fujian has the

---

[11] *Fengshui* forests are usually found in three kinds of places. First, they exist around villages and houses as a symbol of plentitude. Secondly, they exist near graves, especially graves of important figures such as emperors. Finally, these groves survive near temples. In each case, these trees perform a "protective" function. Some *fengshui* forests have a long history. For instance, the grove near the Temple of Confucius includes one hundred thousand trees planted over a 2,400-year period.

highest forest coverage rate in China, the coverage area of natural forest is decreasing day by day. In the future, we may only see natural forest in nature preserves."

This goes far beyond aesthetics and vague preferences for regarding natural forests as superior. According to Yang Yusheng, Professor at Fujian Normal University, the increasing frequency of floods and mudslides in the northern part of Fujian is the direct result natural forest replacement. Insufficient attention has been paid to the consequences of this policy. Impacts on local ecology and biodiversity have obviously been left aside. Butlong-term economic impacts have been overlooked as well. Land afflicted with regular mudslides and flooding will erode productivity and almost certainly produce more poverty.

The aforementioned three examples demonstrate that natural forest destruction is one of the most acute environmental problems in contemporary China. Even as forest coverage expands in some places, the quality and durability of these forests erode.

Certainly, a long history is at work here, as the Chinese have worked the land, including forests, for thousands of years. Even in more recent times, whole tracts of land could be transferred to industrial interests at the stroke of an official's pen. FRR advocates hoped to do away with such flagrant use of administrative power by diffusing control of forestlands and by putting those with a long-term commitment to forest conditions in charge of forest management. Unfortunately, even this system proved vulnerable, as the above examples demonstrate. Now, with many individuals holding the rights to develop forest resources, we may have the same race, only with more runners, and therefore a more complicated management system.

III. *Concerns over Forest Reform in Yichun, Heilongjiang Province*

In June 2006, China's State Forestry Administration approved a pilot FRR project in Yichun. This marks the first case of a formerly state-owned forest entering the new system; to date only collective forests had been subject to forest-rights reform. However, the goals remain the same: increasing economic gains from forest exploitation; improving the ecological function of forests; and moving the broader economic-reform process forward.

Fifteen tree farms in Yichun participate in the reform process. Contracts to use forest resources include various restrictions: maintaining the basic nature and characteristics of forests; continuing pre-reform use patterns, but with refinements; and limiting leases to fifty years. Other than this, those drawing their livelihood from forests receive management rights over and ownership of the timber. Firms win the right to access and use these resources through auctions, sealed bids, or negotiation with local forestry bureaus. Logging methods and amounts must meet legal standards.

As soon as the contracts had been completed, workers threw themselves into afforestation efforts. Therefore, woodland area is increasing instead of decreasing.

Yichun's forests were the first to undergo industrial exploitation after the Revolution. This region of northeast China and Inner Mongolia has been heavily exploited since this process started. Forest reserves have decreased from 428 million m^3 to two hundred and ten million m^3, a 51 percent decline since 1949. Mature forest reserves have decreased from 320,000,000 m^3 to 6,100,000 m^3, a colossal 98 percent decline. Even with massive harvests, lumbering companies have lost money for years. However, recent conditions have improved as more natural forests have come under protection, but these have not addressed the underlying resource crisis and remaining financial difficulties.

Over-logging has been so severe for so long that twelve of the sixteen forestry bureaus in Yichun have no forests to exploit. The other four face a similar future. Payment to companies working forests has been delayed or remains in arrears. The farms themselves carry large debts, some topping the one hundred million RMB mark.

Three phenomena have driven these conditions: large sell-offs of forest assets; the rapid accumulation of wealth among profiteers engaged in lumber speculation; and the emergence of aggressive groups involved in illegal logging.[12]

FRR should put an end to preferential treatment and corruption in forestland management. By engaging more participants, a more democratic decision-making system can take hold. With better representation, more persons should benefit. The expectation is that Yichun would prosper economically and ecologically. Government should provide logistical, legal, and financial support to develop the reforms.

[12] Translator's note: A characteristically aggressive group is the Putian Band from Fujian Province. To some degree they have controlled the forest industry in Yichun, boosting output such that it is now the world's leader in the export of wooden floorboards.

The cold, hard reality does not fit the theory at all. Impoverished workers willingly sell off their rights to outsiders for virtually any amount that looks as if it will bring immediate economic gain. Stressed local banks do little to support locals in blunting this trend.

This process may ramp up in scale with the 2007 opening of the Forest Rights Exchange Center. Groups like the Putian Band have watched this process with keen interest because it would allow them to purchase land rights *en masse*. This will further concentrate their economic and political power in the area.

Businessmen, such as those with Putian, operate under Machiavellian principles. They ignore the environment and environmental regulations, except to challenge or complain about them. They bribe those they can and deceive the rest. They carefully identify bottlenecks and checkpoints designed to regulate, and then transform these into "green channels" through which timber flows into markets and money flows into the pockets of those supposedly minding the checkpoints. While money is made, natural forests suffer, and locals barely eke out a living from the land. But in the minds of industrialists, and those complicit in advancing their efforts, such outcomes are justified under the aegis of "reform," "development," "modernization," and "growth."

IV. *Suggestions and Countermeasures*

At a February 27, 2006, press conference, Jia Zhibang of the State Forestry Administration said:

> During the Eleventh Five-Year Plan (2005–2010), China will adopt a comprehensive forestry strategy aimed at ecological protection; developing a system that balances forest ecology and industry is paramount. The following goals will be met during the Plan's duration: forest coverage will reach 20%, up from the current 18.21%. Forest reserves will reach 13.2 billion m^3. All public forests... will receive effective protection. The total revenues from forest exploitation will reach 1,200 billion RMB."[13]

[13] Jia also pointed out that, "...50% of natural wetlands, and 90% of wildlife under first class state protection... will receive protection." These statements come from a broadcast of "The Eleventh Five-Year Plan on Forestry Development Is Out in China: Three Forest Right Reforms Take Effect This Year," CCTV International, February 27, 2006, http://www.cctv.com.

Elsewhere in this proclamation, Jia stresses five principles that the Chinese government should stick to while implementing FRR on collectively held land. First, the government should exercise different kinds of supervision over different types of forests, applying the most appropriate means and measures in each given situation. Public forests should remain off-limits for commercial exploitation. As compensation for this restriction, greater latitude could be granted to those working commercial forests. This could better balance ecological and utilization goals. Second, the government should respect the will of the farmers and locals working forestland, and place their interests ahead of those of other interest groups. Third, the government should handle forest affairs openly, impartially, and fairly. The content, procedures, and results of reform should be open to the public, to ensure a role for public participation, comment, and oversight. Fourth, the government should tailor guidelines to best fit particular social, cultural, and economic conditions and needs. Fifth, sustainability and ecological stability should move to the middle of the policymaking philosophy and ethos.[14]

Many more issues need to be addressed. Finding the right balance between the flexible approaches advocated here and a more systematic, standardized approach requires greater exploration and development.

A. *Raising Awareness of Natural Forest Protection*

As noted, forest cover is slated to reach 20 percent of China's land area by 2010. Jiang Jiasi, dean of Peking University's Environment College, believes that expanding coverage to 30 percent would resolve numerous environmental problems. While this may be the case, an additional condition must be considered within this proposal: protecting natural forests.

Although China has formulated strict policies on paper regarding natural forest protection, in fact these forests remain in jeopardy. The profit incentive prods local governments and farmers to cull natural forests, replacing them with fast-growing trees or fruit trees. Market forces also have a macro effect on protection efforts: as demand for timber continues to soar, so does pressure on existing natural forests. High-quality wood has many uses, and almost all of them fit into expanding

[14] Qin Jiaofeng, "Initiating Forest Rights Reform: An Interview with Jia Zhibang," *China Comment*, 16 (2006).

market segments: newsprint; office and copier paper; tissues; furniture; floor boards; disposable chopsticks;[15] and so forth.

Aggravating the input problem is the overall inefficiency of Chinese production systems. For instance, paper mills look almost exclusively to wood suppliers for making paper, rather than to other inputs such as straw. The wood most often used, and officially approved of and supported by the China Paper Association, is eucalyptus, which is one of the most damaging to local ecological systems. These trends certainly point to a troubling future for China's remaining natural forests.

Successfully implementing and expanding such efforts will require a combination of rhetorical and economic strategies. First, biodiversity and native species should be privileged over approaches that degrade either. These policies should apply in those cases where existing timber is harvested and replaced. In such cases, replacement should proceed with surgical care rather than with chainsaw recklessness.

People living in and working in forests must better understand that forest protection does not conflict with, and in many cases can enhance, economic benefits flowing from forest use. This will require more than just persuasion, however. The traditional agricultural population should be reduced; they should receive skills training in other areas—creative or high-tech work, tourism services, and organic, cooperative, and high-margin agriculture are viable alternatives—to ensure their economic well being.

B. *Expanding the Farms-to-Forests Program*

Economic benefits realized through FRR have brought on a reassessment of farmland-to-forest (F2F) policies. Some quarters have even advocated rolling back the policy, saying that forest reform provides sufficient benefits to rural areas already. On the other hand, some experts have suggested expanding the thrust of this campaign to include lakes and grasslands, with an eye toward nurturing permanent vegetation in larger areas. Even within its limited tenure, the F2F program has achieved positive results. Given that more forestland is needed and that

[15] Translator's note: China plays a pivotal role in the supply of these goods for markets around the world. For instance, Japan imports 90 percent of its disposable chopsticks from China.

natural forests need ongoing and expanded protection, both programs should continue.

C. *The Necessity of Public Participation and Oversight*

Given the complexity of reforms, environmental NGOs (ENGOs) have a crucial role to play. ENGOs can organize public participation, monitor government activities, provide information to the media, and carry out independent research and data collection on conditions and impacts of environmental policies and their implementation. These efforts can be directed toward reports as well as advocacy campaigns.

While laws on these matters are clear and straightforward, implementation has been murky and convoluted. By focusing public interest, local government will pay more attention to environmental protection while more effectively balancing such concerns with economic development.

D. *Reform Beneficiaries Should Take and Bear Greater Responsibility for Their Actions*

Too many feeding at the public trough have failed to act property or shoulder their fair share of the load in improving environmental conditions. For instance, farmers granted rights to forestland must think carefully about how to use that land and be effective stewards for its management. Moreover, failure to do so should result in their bearing some liability for harm to the land and its resources.

Such efforts can be bolstered through legal means. Legislation on ecological restoration, public interest lawsuits on environmental matters, creation and implementation of environmental insurance, and compensation schemes for those physically, economically, or culturally harmed through environment destruction require formulation and circulation. The underlying goal should be to internalize environmental costs typically counted as externalities. Thus, those who inflict harm on the environment will feel the pain and burden of doing so in economic terms. Clearer guidelines regarding legal repercussions need development; the most egregious environmental crimes may require non-monetary punishment, such as imprisonment.

Legal reform in environmental matters must surely include the development of public interest litigation. This tool has had a powerful effect on markets and cultural beliefs in developed countries and could offer similar benefits in China. However, the Chinese legal system offers virtually no support for public interest litigation. This deprives advocates of natural forests and the environment at large of a potentially powerful tool in their continued protection efforts. We may well have a long way to go until this and other supporting reforms take hold in China.

CHAPTER NINETEEN

CHINESE WOOD PRODUCTS TRADE AND THE ILLEGAL TIMBER TRADE[1]

Tamara Stark, Shi Pengxiang, and Cheng Yun

In China, recent years have witnessed soaring consumption and trade of wood products.[2] The impact of booming demand for wood products on forests has attracted attention from the international community and environmental protection organizations. Some NGO reports about global supply chains reliant on Chinese wood products have recently garnered attention from the international media. A BBC article entitled "China Furniture Destroys Forests," cited Forest Trends's[3] report that forestland is being seriously damaged due to China's voracious demand for wood products. Another article from the *Financial Times* reported that because of China's hunger for wood, and insufficient supervision of commercial operations and import guidelines, forests in Indonesia and Papua New Guinea have been seriously damaged. This article was based on a Greenpeace report entitled "Sharing the Blame—Global Consumption and China's Role in Ancient Forest Destruction." Both reports point to the final link in this industrial chain: end consumers in the European Union (EU), the United States and Japan, noting that these countries must do more to stem the flow of illegally harvested wood.

The remaining ancient forests in the world face life or death conditions. Huge demands from developed countries propel surging growth of the timber and furniture industry in China. To meet demands, wood processing companies, such as furniture and paper manufacturers, import increasing volumes of wood. This, in turn, encourages more and more illegal logging and over-logging. Both China and the international community now understand that the time has come to objectively evaluate the impact of China's timber trade on global forests, and based on this evaluation, to implement effective forest protection measures.

Key words: wood products, illegal timber trade, forest certification

[1] Translator's note: This article, except Part I, is based on the Greenpeace report entitled "Sharing the Blame—Global Consumption and China's Role in Ancient Forest Destruction." All related statistics and charts are quoted from that report.

[2] Translator's note: Unless defined otherwise, "wood products" in this article refer to all products manufactured from wood, wood pulp, and paper.

[3] Translator's note: Forest Trends is an international environmental NGO involved in forest protection and improved management. More information can be found at: http://www.forest-trends.org/whoweare/mission.htm.

I. *Analyzing China's Wood Market*

China is the world's second-biggest importer of wood products. The country imports more raw wood and tropical wood products than any other country. Much of this is raw timber or wood used as inputs in various industrial processes in Chinese factories. Even though it imports so much, this market is really export-driven. That is, export growth exceeds the growth rate of domestic consumption; much of what comes in goes out again. In short, China serves as the "world's factory" for plywood, furniture, paper, and other processed and finished wood products.

China's vast industrial capacity and its remarkably low production costs exert considerable impact on the global wood-products market. For example, developed countries no longer import timber directly from forest-rich countries, but rather purchase processed wood products from China. At the same time, these countries demand more processed and finished wood products as prices fall in response to the cost benefits of Chinese production. This has increased Chinese timber demands. Part of this demand is met with timber from areas where over-logging and illegal logging are practiced. Wood products from all these sources eventually leave China as processed and finished goods destined for Europe, the United States, or Japan. China has become the pivotal waystation and central processing center for the global timber trade.

In 1998, the Chinese government launched the Natural Forest Protection Project (NFPP). The Project sets strict limits on tree harvests, which marks a significant change in China's forestry policy. To satisfy demand while observing these restrictions, wood processing enterprises looked abroad for timber. So while 1998 marked a new direction in domestic forestry management, it also marked the year when China became one of the world's biggest importers and consumers of timber. This makes China as dependant on global forest supplies for its wood production and processing as the world has become on China's wood product exports.

A. *Analyzing China's Wood Consumption and Trade*

China's growing demand for wood inputs pressures global forest resources. According to Figure 19.1, China's demand for wood increased by 70 percent from 1995 to 2004. World wood demand increased by

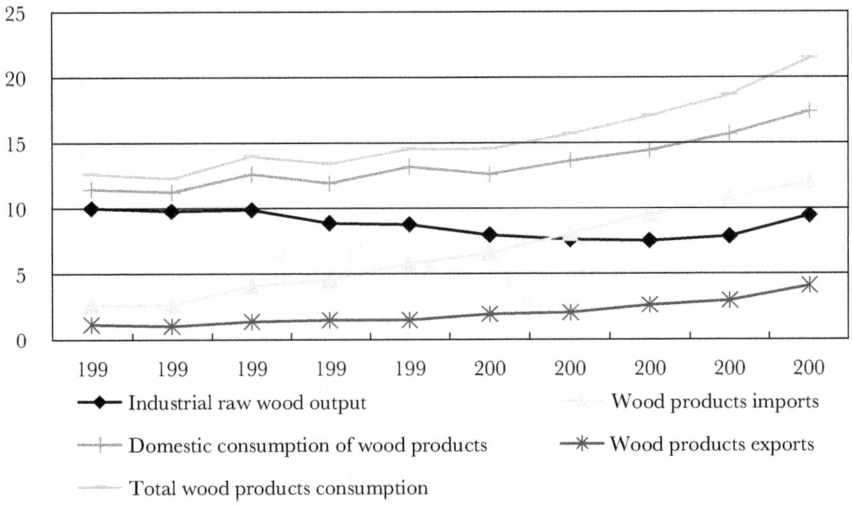

Figure 19.1: Production, consumption, import and export volumes of China's industrial wood products between 1995 and 2004 (in million m^3 RWE)

108,000,000 cubic meters[4] in these years; China's demand shot up by 59,670,000 m^3 RWE (Round Wood Equivalent), accounting for 55 percent of the world increase.

Because the global appetite for timber has grown so intense, China, and its downstream customers in rich and developing countries alike, often turns a blind eye to the provenance of wood supplies. China takes much of the direct blame for this, since it directly imports wood from over-logged and illegal sources. However, rich countries exercise a practiced ignorance in this process. They certainly are complicit in allowing forest mismanagement to continue.

Moreover, China's per capita consumption remains below the world average, and far below the levels in developed countries. For instance, domestic consumption of finished paper products in China is only 12 percent of the United States's demand, and 15 percent Japan's. Laying so much of the blame for forestry mismanagement on China's doorstep hardly seems just or reasonable.

[4] FAO Forestry Data. http://www.faostat.fao.org/faostat/collections?subset=forestry.

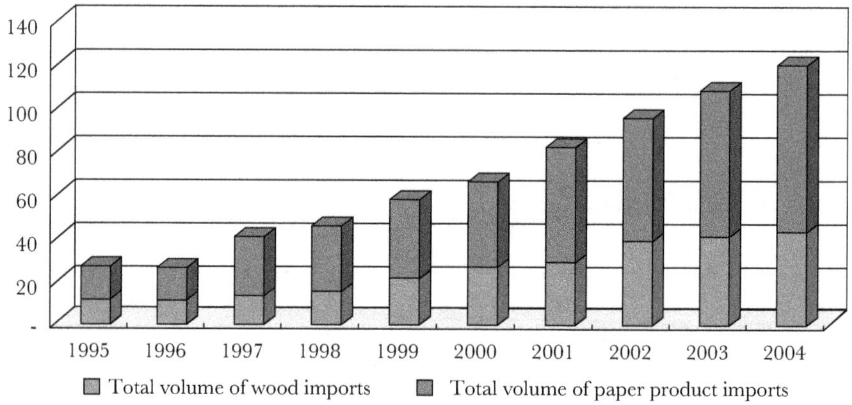

Figure 19.2: Wood imports, 1995–2004 (in million m^3 RWE)

B. *Analyzing China's Wood Imports*

Due to strict limits on domestic logging, China's wood-processing industries are importing enormous amounts of wood. Imports reached one hundred and twenty million cubic m^3 RWE in 2004 (see Figure 19.2), roughly 4.5 times import volume in 1995. China is now only second to the United States in raw wood and wood pulp imports. If industrial demands continue to increase while domestic supply limitations remain in place, import volume will continue rising.

Government policy has helped to drive the expansion of imports. A key move came in January 1999, when officials set tariffs on wood imports to zero. Almost all wood products qualify: raw wood; sawed wood; fuel wood; wood chips; wood pulp; and waste paper. This has helped to attract various kinds of wood-processing operations. Other policies have bolstered this process as well, including a ban on raw wood exportation, and strict limits on sawed-wood exports. The emphasis has moved to the promotion of relatively high value-added goods. This should prod Chinese producers to move out of the high-input, low-margin commodity-based segment of the timber and wood products market.

C. *Analyzing China's Wood Exports*

From 1995 to 2004, China's wood exports grew by 350 percent, putting China near the top of global processed-wood-products exporters.

It is now the largest exporter of plywood and is likely to become the number one furniture exporter, taking the place of Italy.

This rapid growth has been achieved as a result of governmental policies as well as ready access to low-cost inputs, such as raw materials and labor. Government subsidies offered to makers of high-value-added processed wood products have been a particularly potent policy. For instance, pressboard, plywood, wood furniture, and some paper products garner payments of up to 13 percent of the export price.[5]

Figure 19.3 shows that wood product exports doubled from 2000 to 2004. Even after establishing a large baseline volume, growth has remained strong: 2004's export growth rate reached an historic high of 35 percent. These exports typically go to rich markets such as those of the United States, Japan, South Korea, Hong Kong, Taiwan, Canada, and EU countries. The United States leads this pack, buying 25 percent of China's exports. The United States also dominates the list of wood furniture importers, buying about half of China's output, with the remainder going to Hong Kong, Japan, the EU, Australia, Canada, and others.

The leading buyers of China's paper products are much the same: Hong Kong, the United States, Japan, Taiwan, Australia, South Korea, and the EU. Hong Kong tops the list, partly to supply its re-export businesses.

As with much of China's industrial sector, low production costs drive investment from foreign firms. Many such companies establish production facilities in China to serve their home market, as well as other markets. Foreign-held firms in Guangdong Province exported more than two billion USD worth of furniture, accounting for 57 percent of the total goods exports from the province during the first eight months of 2005.[6] These manufacturers prefer timber from aboard. For example, some expensive woods from the United States go into

[5] Translator's note: Such policies may have triggered a response from importers, however. For instance, the EU instituted an anti-dumping policy against some Chinese-made plywood products in 2004. In spite of this, EU statistics show that from January to November 2005, the EU's plywood imports from China increased by 80 percent. However, by November 2005, import activity had slowed to "only" a 65 perecent year-on-year increase. See Eurostat: http://www.europa.eu.int/comm/eurostat/.

[6] Data Office of Guangdong Branch of China Council for the Promotion of International Trade, 2005.

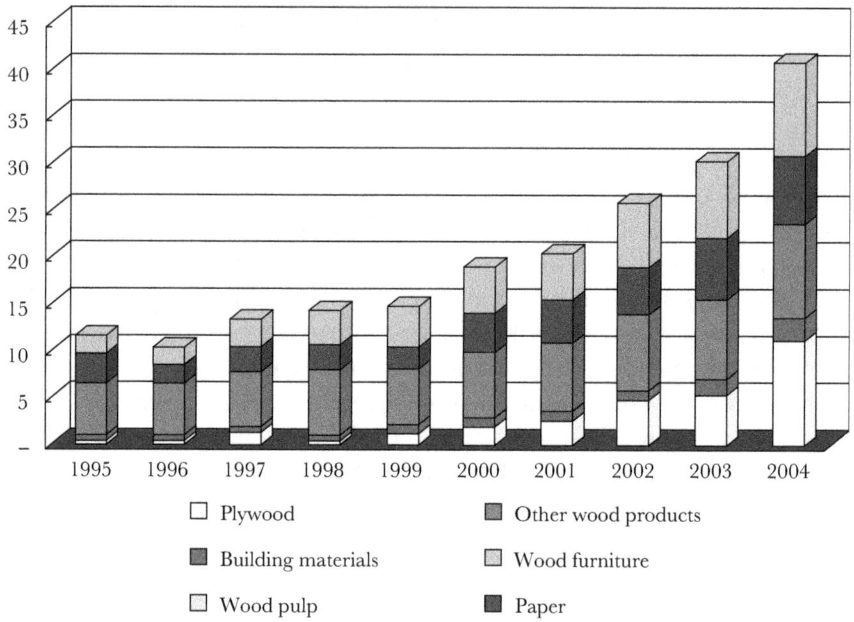

Figure 19.3: Wood product exports, 1995–2004 (in million m^3 RWE)

furniture specifically targeting more upscale markets, such as that of the United States.

D. *Over-Logging and Illegal Logging*

Comprehensive and accurate data on illegal activities is hard, if not impossible, to find or compile. Illegal logging and trading is no exception. However, international organizations have constructed estimates based on trade volume, prices, and documentation of legal and illegal logging. The World Bank estimates that 40 percent of the wood materials imported from the easternmost part of Russia have been harvested illegally.[7] According to the United Nations Food and Agriculture Organization (FAO), logging and wood poaching places forests in imminent danger. Although data regarding the trade of illegal timber cannot be

[7] J. Blazer et al., paper prepared for the Ministerial Conference, November 22–25, 2005, at St. Petersburg, Russia, Forest Law Enforcement and Governance (FLEG) in Eastern Europe and Northern Asia (ENA).

considered wholly accurate, information suggests that China may be the world's number-one destination for such products.[8]

The FAO goes on to note that most countries supplying wood materials to China face diminishing forests, and with them diminishing timber supplies. The Chinese government and Chinese forestry experts recognize this. Given that imports probably cannot meet the country's growing input needs, China has engaged itself in an ambitious afforestation campaign.[9] This approach may also reach its limits. For instance, given China's soil and water limitations, many farmers prefer growing commercial crops to timber.

Clearly, China shares some of the responsibility for stoking this process, which has earned it many critics. Unfortunately, its continued voracious appetite for wood of any kind may promote illegal logging.

E. *Government Forestry Commitments*

The Chinese government faces two serious questions regarding wood material needs: Can suppliers provide China with sufficient, continuously delivered supplies? What volume of imported timber comes from forests where over-logging and poaching continue? Sustainability ties together both questions. China must meet reasonable demands, but not at the expense of long-term supplies.

As a leading developing country, China has demonstrated a strong willingness to participate in the international community's environmental protection efforts in recent years. There is a growing awareness among the Chinese that the country should not copy the profligate high-consumption lifestyles and economic models of developed countries, because doing so may well deplete global resources. This awareness extends to the government's policies regarding forest management. China has become a signatory to bilateral and multilateral agreements to show its readiness to crack down on illegal logging and trading. The government also joined commitments affecting forestry through a series of documents, such as Forest Law Enforcement and Governance (FLEG), the United Nations Convention on Biological Diversity, the

[8] For example, see the United States Department of Agriculture (USDA) report that regards China as one of the leading destinations and processors of illegal timber. Foreign Agricultural Services (FAS), USDA, "People's Republic of China Solid Wood Products," (2005), http://www.fas.usda.gov/gainfiles/200507/146130349.pdf.

[9] Translator's note: See Chapter 17 of this volume.

St. Petersburg Ministerial Declaration, the Memorandum of Understanding between China and Indonesia, the China-EU Summit Joint Statement, and the Sino-Russian Joint Communiqué.

Given such efforts, developed countries have little justification for laying so much blame on China for importing timber of questionable provenance and fueling forest destruction. Again, we reiterate: final buyers—generally these same developed countries heaping blame on China for its role in illegal logging and deforestation—import illegal timber. It simply comes into their countries as finished wood products, but their appetites and callousness fuel illegal logging and deforestation at least as much as Chinese actions. There is plenty of blame to go around; all those responsible must do more to make improvements to the global wood trade.

II. *Wood Inputs and Old-Growth Forests*[10]

The FAO estimates that global forests lose 7,300,000 hectares annually.[11] Old-growth forests face particularly acute risks according to the World Research Institute (WRI). The WRI claims that commercial logging is the single-greatest threat to such forests. Logging threatens 70 percent of forests on the verge of extinction globally.[12] Old growth forests have dwindled to near-nothingness. Even northern coniferous forests and southern tropical forests[13] face mounting threats from large-scale industrial operations. If these trends continue, we will undermine forest ecology, and perhaps economic systems as well.

A. *Threats to Old-Growth Forests*

Old-growth forests provide habitat for two-thirds of the world's land-based animal and plant species, and play a vital role in maintaining climatic stability and ecological balance. They are also home to tens

[10] Translator's note: Readers from various countries may know "old growth forest" under another name: primary, ancient, primeval, or virgin forest. The terms "frontier forest" and "ancient woodland" are also used.

[11] The United Nations Food and Agriculture Organization (FAO), "Global Forest Resources Assessment" (2005), www.fao.org/forestry/fra2005.

[12] World Research Institute. 1997.

[13] Translator's note: Even though damage to tropical forests has been widespread, almost all of it has occurred since 1949.

of millions of indigenous peoples who have lived in the forests for generations and depend on them for their livelihood and way of life. In the late 1990s, satellite images made available through the World Resources Institute showed the scarcity of virgin forests; most of these face ongoing dangers from industrialization. Unsurprisingly, these primeval forests face a grave situation. Greenpeace staffers, international scientists, and cartographers have also assessed these areas using satellite imagery. They found that old-growth forests cover less than 10 percent of the Earth's land area and eighty-two out of 148 countries lying within forested zones have lost virtually all their intact old-growth forests. Younger forests are rapidly disappearing as well. Poor logging practices, over-logging, tree theft, and poaching are the main causes of this outcome.

B. *Wood Imports from the "Paradise Forests of the Asia Pacific"*[14]

Indonesia and Papua New Guinea, home to some of the largest tracts in the Paradise Forests of the Asia Pacific, are losing old growth forests faster than any place on Earth.[15] Illegal logging and over-logging are widely practiced in these forests. Much of this wood, as well as other timber from the Pacific, is exported to China, which is the leading importer of wood from Indonesia[16] and Papua New Guinea.

C. *The Scale and Impact of the Illegal Timber Trade*

By using import figures from Chinese Customs and export figures from various wood suppliers, estimates of the illegal timber trade can be estimated. Chinese Customs shows that 2,810,000 cubic m^3 of raw timber entered the country from Indonesia and Malaysia; 1,660,000 m^3 of this could not be accounted for in the corresponding export records of the supplier countries. In other words, 59 percent of China's raw wood imports from these two countries could not be accurately sourced. It

[14] Translator's note: "Paradise Forests of Asia" is an assignation granted by Greenpeace to ecologically important forests, especially large old growth forests containing rare and diverse species of flora and fauna.

[15] Translator's note: A map of these forests can be seen here: http://www.greenpeace.org/international/campaigns/forests/asia-pacific.

[16] Translator's note: Timber from Indonesia is often disguised as wood of Malaysian origin in order to sell the wood to Chinese importers. Such clever schemes need to be defeated through better product tracking methods, among other actions.

is likely that a significant proportion, if not all, of it came from illegal logging. Note that some recorded exports from source countries may well include illegally harvested wood. Furthermore, the above figures only account for wood from two countries supplying timber and wood supplies to China. In short, the scale of illegally harvested wood entering China may be much larger than these figure suggest.

In recent years, China's major wood materials importers and manufactures have become increasingly concerned regarding the source of wood purchases. At the same time, more and more overseas buyers ask Chinese producers to provide documents that certify the legitimacy and sustainability of timber supplies.[17] Both forces place market pressure on Chinese operators, since their processed and finished goods typically go to developed countries that have to confront stronger environmental awareness among consumers and higher environmental standards from regulators. Such pressures bring up relatively new and largely unconsidered issues for Chinese firms, and these pressures seem unlikely to diminish. With its entry into the WTO, China must move swiftly to implement forest certification systems that harmonize with those practiced in more developed markets. The future of its multi-billion-renminbi wood-processing industry—and more importantly, forests worldwide—depends on it.

D. *China's Choices*

China's policies and actions will affect forests worldwide, but will have major impacts on the Paradise Forests and their environs. By meaningfully and sincerely tackling the problem of importing illegally harvested wood and/or wood from over-logged land, China can contribute significantly to the survival of the Paradise Forests. Implementing such measures will also affect major wood materials suppliers around the world, such as Russia, Burma (Myanmar), Brazil, Gabon, and Congo. The future of their forests will depend on China's stance regarding unsustainable logging practices.

[17] USDA FAS. 2005.

III. *The Effects of Certification on the Timber Market*

Fortunately, the international community has begun to combat illegal and dangerously destructive logging and its associated trade. International media frequently issue alerts and reports about the condition of global forests, including the rising threat of over logging and poaching. Environmentalists from all over the world also continue to exert an important influence on forest preservation. In particular, they call on the international community to be more cognizant of the risks facing forests, to resist importing or using illegal timber products, and to work collectively toward forest protection. All nations should work to assure that forests continue to provide ecological, social, and economic benefits. Below, we discuss forest certification and its impact on this process.

A. *Forest Certification*

In the early 1990s, various organizations and individuals successfully linked sustainable forest management to policies and actions affecting global timber markets. At the individual level, environmentally aware consumers focused purchases on timber products from well-managed forests. To support and expand this process, responsible companies, nongovernmental organizations (NGOs), and advocates developed and launched forest-certification campaigns. Products made from wood culled from well-managed forests are marked with special tags so that consumers can readily identify these products and support responsible forest-management practices. Forest certification consists of Forest Management Certification (FM)[18] and Chain of Custody Certification (CoC).[19]

As a subsidiary step, these forces pushed for greater information gathering and transparency regarding such points as establishing a given batch of timber's provenance. Certification has quickly become an effective way to adjust markets, protect forest resources, and balance environmental and economic benefits.

[18] Translator's note: FM is described in more detail here: http://www.rainforest-alliance.org/forestry.cfm?id=management_certification.

[19] Translator's note: CoC is described in more detail here: http://www.rainforest-alliance.org/forestry.cfm?id=custody_certification.

Forest certification aims to improve forest management, to foster sustainable development, and to guarantee market access for responsibly-grown and—harvested timber. This independent assessment system helps to link green consumers with wood products suppliers that meet certification standards, helping to financially support those firms with a verifiable record of sustainable forest management.

B. *The Influence of Forest Certification on China's Wood-Products Trade*

Heated discussions promoting forest certification abound and have vastly expanded these programs. As of January 2006, 68,125,000 hectares of forest in sixty-six countries had been certified by the Forest Stewardship Council (FSC); 4,350 enterprises in seventy-three countries earned the Chain of Custody Certificate. Other large, forested areas have been endorsed by other certification systems. Clearly, the system continues to exert an expanding impact on the global wood-products trade.[20]

C. *Immediate Concerns and Problems Regarding Certification*

Forest certification in China is still in its early stages. In July 1999, at an international symposium on forest certification held by the State Forestry Administration of China and the World Bank/World Wildlife Fund Alliance, the concept of forest certification was introduced in China for the first time. In July 2001, the State Forestry Administration established the China Forest Certification Work Group, marking the official launch of forest certification work in China.[21]

Two forces have helped to drive the adoption of certification in China by increasing international pressure, largely through market forces, and increasing commitments made by the Chinese government to establish and encourage sustainable forest management practices. By March 2006, 137 timber-processing enterprises had acquired the Chain of Custody Certificate issued by the FSC, most of them foreign-owned or Sino-foreign joint ventures producing specialized, small-production-run furniture. FM certification also expands apace: 439,000 hectares

---

[20] Yan Shuai, Ning Yanjie, and He Xiaoqi, "Three Basic Activities for Forest Certification," *Journal of Beijing Forestry University (Social Sciences)* 2, no. 4 (2003): 14–17.

[21] "Promoting Responsible Forest Management with All Your Support," Forest protection project run in cooperation between WWF and IKEA Group-China (2006).

of forest managed by four companies have received certification from the FSC.

Even with these victories, the process has a long way to go. Various impediments still exist: high certification costs; frictions in applying international certification systems in China; the lack of a domestic certification body; and perhaps most of all, ignorance of the certification system and its processes.

D. *The Prospects for Certification*

The recent 2005–2006 Market Report by the United Nations Economic Commission for Europe (UNECE) and the FAO shows that forest certification has established genuine traction.[22] This innovative approach for promoting sustainable forest management has the potential to change traditional forest management modalities, and thus have a huge influence on the world's wood-products trade.

IV. *Reflections on China's Place in the Global Forestry Crisis*

If China increased its per capita paper consumption to that of the United States—moving from 36 kg to 301 kg in the process[23]—it would consume nearly 1.6 billion m^3 more timber, or the equivalent of the world's annual timber harvest.[15] If developed countries do not curb their wood consumption, and similarly, if China does not slow its growing appetite for wood and wood products, future generations will live on a planet denuded of old-growth forests.

China's growing demand for wood-products inputs fosters two key problems: illegal and irresponsible logging and its associated trade, and unsustainable consumption levels.[24] China, and its downstream consumers, must come to grips with the problem of illegal logging and over-logging. This heightened awareness must give rise to decisive and concerted action to remedy the problem. Moreover, because this

[22] "An Annual UN Report—Forest Certification is Becoming a Trend." *China Green Times*, November 8, 2006.

[23] FAO Forestry Data. faostat.fao.org/faostat/collections?subset=forestry.

[24] Translator's note: Because of China's central role in the world's wood and wood-products markets, these problems also face all producers of timber and timber-based goods.

is an international problem, it must be addressed through international institutions and with cooperation and coordination that includes China and other nation-states. At present, though, no international mechanism exists to coordinate such an agenda. However, some elements of a solution exist, such as certification programs.

China's—and the world's—growing demand for wood-based products continues to rise quickly. Forests may not be able to sustain current levels of consumption, much less some of the projected demand levels. Conservation, alternative materials, and other means of production must develop to protect declining forests. Among other actions, China can advance the use of substitute materials, such as straw, strengthen timber quarantines, better control invasive species in forests, and encourage the growth of rare tree species in or near manmade forests.

Finally, the world's governments, enterprises, and consumers must realize that forest resources are limited. Among other things, this means that the economic model and lifestyle in place in developed countries may not work for other countries and seems unlikely to work for the planet as a whole. In fact, these things may have outlived their usefulness for developed countries as well. The goal should be to move toward durability in forest management and timber markets, as part of a broader shift toward new forms of economic production, distribution, and consumption. Together we must search for a new model that can bring sustained economic, social, and ecological benefits.

APPENDIX

WHAT IS A QUALIFIED CERTIFICATION FOR WOOD?

The most effective way to ensure that wood has been legally and responsibly harvested is to establish the provenance of that wood and to follow it through the entire supply chain. Widely-recognized forest-quality certification and market supervision certification accomplish these tasks. Certification has been established by the following organizations:

- Canadian Standards Association (CSA)
- Forest Stewardship Council (FSC)
- Malaysian Timber Certificate Council (MTCC)
  - Programme for the Endorsement of Forest Certification (PEFC)
- Sustainable Forestry Initiative (SFI)

To date, the FSC has authorized the following institutions to grant certifications in China:

SGC, www.sgs.com
Smartwood, www.smartwood.org
GFA Consultation Group, www.gfa-grounp.de

In lieu of certification, a company can have its supply chain audited and present the results of that audit to buyers, regulators, or interested parties. The following documents should be examined for validity:

- Original business letters, sales invoices, transportation documents, and receipts
- Operating licenses for logging operations, including management schemes approved by supervising regulators
- Certification of use tax payments

While documents establishing legal operations are important, it is not sufficient in the fight against over-logging, theft, and poaching.

Certification standardizes processes and benchmarks. Moreover, centralizing and/or coordinating data collection through a few bodies improves information tracking and sharing, which in turn helps with policymaking and enforcement.

PART FIVE

APPENDIX

APPENDIX

ANNUAL INDEXES: ENVIRONMENTAL TRENDS

This appendix to *The China Environment Yearbook, Volume 2* provides a series of brief descriptions of trends at work in environmental conditions, policy, and activism. These data are summarized in figures, tables and graphics below. The data cover an array of areas: air and water pollution, urban environmental conditions, energy consumption, and overall ecological situation.

Even though data are not complete or comprehensive, general tendencies can be perceived from these indexes, especially for the period of the Tenth Five-Year Plan. Future volumes of the yearbook will update these metrics, while adding new data that deepens insights into China's current and figure environmental conditions.[1]

I. *Air Quality*

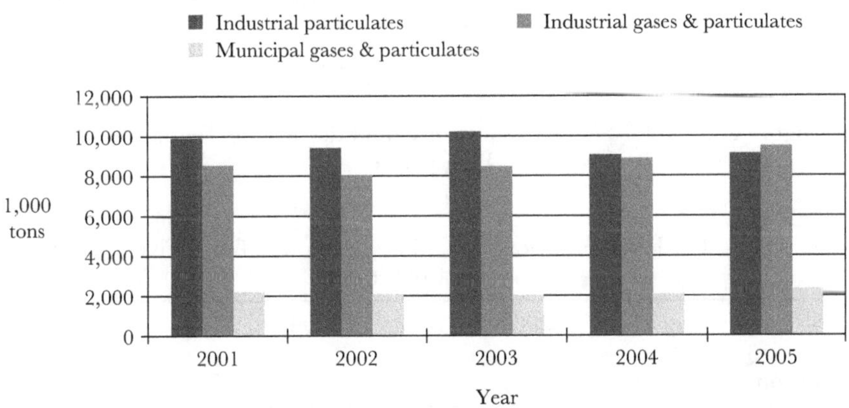

Figure A.1: Total industrial emissions in China during the Tenth Five-Year Plan period

[1] Source: All data are from SEPA unless otherwise indicated.

Table A.1: Total industrial emissions in China during the Tenth Five-Year Plan period (1,000 tons)

Emission \ Year	2001	2002	2003	2004	2005
Industrial particulates	9,906	9,401	10,210	9,048	9,112
Industrial gases & particulates	8,519	8,042	8,462	8,865	9,489
Municipal gases & particulates	2,179	2,085	2,025	2,085	2,336

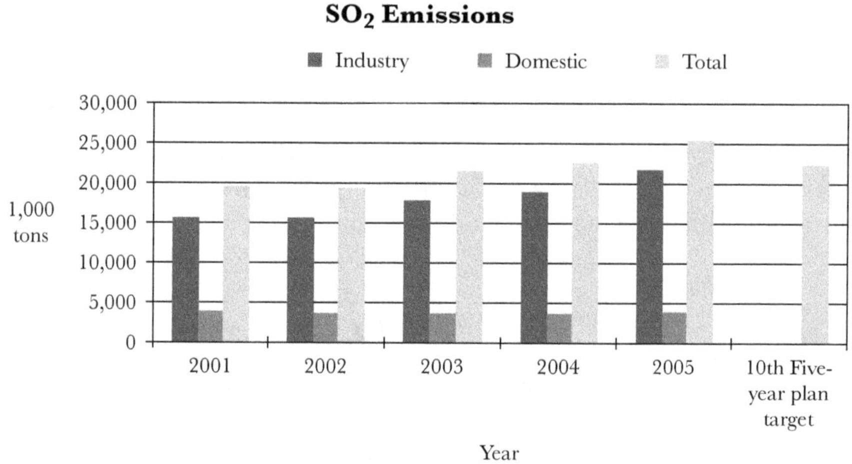

Figure A.2: Total SO_2 emissions in China during the Tenth Five-Year Plan period

Table A.2: Total SO_2 emissions in China during the Tenth Five-Year Plan period (1,000 tons)

Emission \ Year	2001	2002	2003	2004	2005	The 10th Five-year Plan target
Industry	15,666	15,620	17,914	18,914	21,684	–
Municipal	3,812	3,646	3,673	3,635	3,809	–
Total	19,478	19,266	21,587	22,549	25,493	22,490

In accordance with guidelines laid down by the State Environmental Protection Agency (SEPA), we focus on three factors affecting SO_2 targets: (1) Extraordinary increases in energy consumption caused SO_2 emissions to get out of control; (2) The rate at which desulfurization projects have come online has not kept pace with the growth of SO_2 emissions; and (3) Market failure has plagued the development of desulfurization capacity and implementation. In sum, desulfurization programs fell short because of rapid growth in project construction, inept project management, and inadequate maintenance of those projects that did come online, as evidenced by the lower operational efficiency of extant projects and the number of accidents association with these projects.

II. *Water Quality*

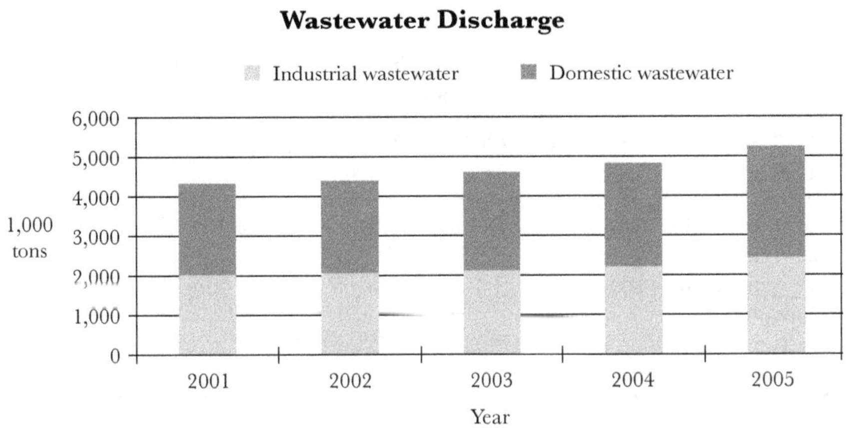

Figure A.3: Total wastewater discharge in China during the Tenth Five-Year Plan period (1,000 tons)

Table A.3: Total amount of wastewater discharge in China during the Tenth Five-Year Plan period

Discharge \ Year	2001	2002	2003	2004	2005
Industrial wastewater	202.6	207.2	212.4	221.1	243.1
Municipal wastewater	230.3	232.3	247.6	261.3	281.4

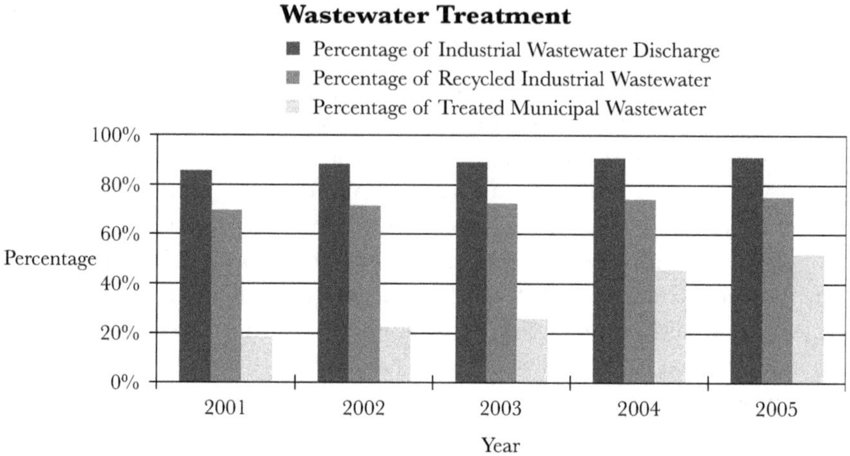

Figure A.4: Wastewater treatment in China during the Tenth Five-Year Plan period

Table A.4: Wastewater standards, recycling, and treatment in China during the Tenth Five-Year Plan period

Percentage \ Year	2001	2002	2003	2004	2005
% of industrial wastewater discharge	85.6%	88.3%	89.2%	90.7%	91.2%
% of recycled industrial wastewater	69.6%	71.5%	72.5%	74.2%	75.1%
% of treated municipal wastewater	18.5%	22.3%	25.8%	45.6%	52%

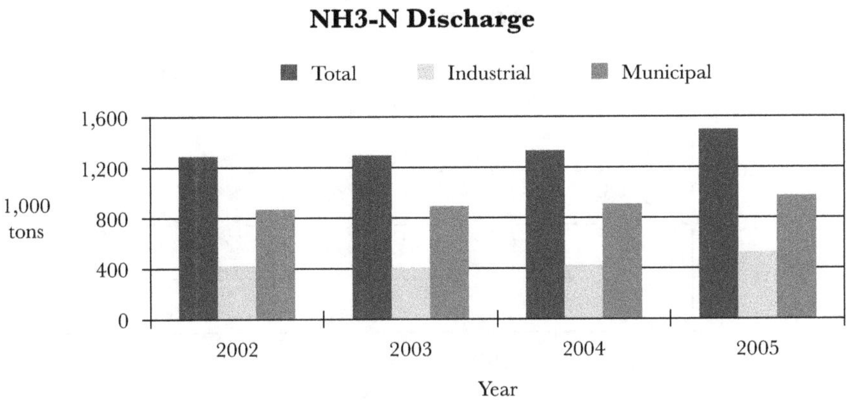

Figure A.5: NH_3-N discharge in China during the Tenth Five-Year Plan period (in 1,000 tons)[2]

Table A.5: NH_3-N emissions in China during the Tenth Five-Year Plan period (1,000 tons)

Discharge \ Year	2002	2003	2004	2005
Total NH_3–N discharge	1,288	1,297	1,330	1,498
Industrial NH_3–N discharge	421	404	422	525
Municipal NH_3–N discharge	867	893	908	973

[2] Ammonia Nitrogen is found in water. In excess it can give rise to eutrophication, which can deprive a given water supply's oxygen. This can have disastrous results for higher life forms such as fish and shellfish. Ammonia Nitrogen generally comes from agricultural runoff.

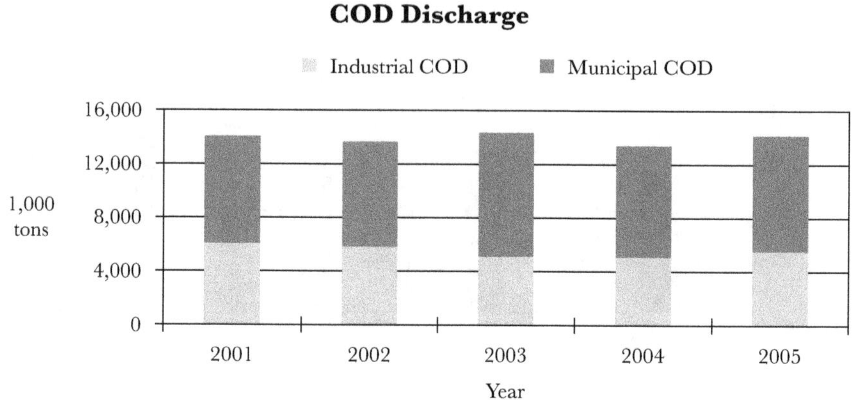

Figure A.6: COD discharge in China during the Tenth Five-Year Plan period (1,000 tons)

Table A.6: COD discharge in China during the Tenth Five-Year Plan period (1,000 tons)

Discharge \ Year	2001	2002	2003	2004	2005
Industrial COD discharge	6,075	5,840	5,119	5,097	5,548
Municipal COD discharge	7,990	7,825	9,217	8,295	8,594

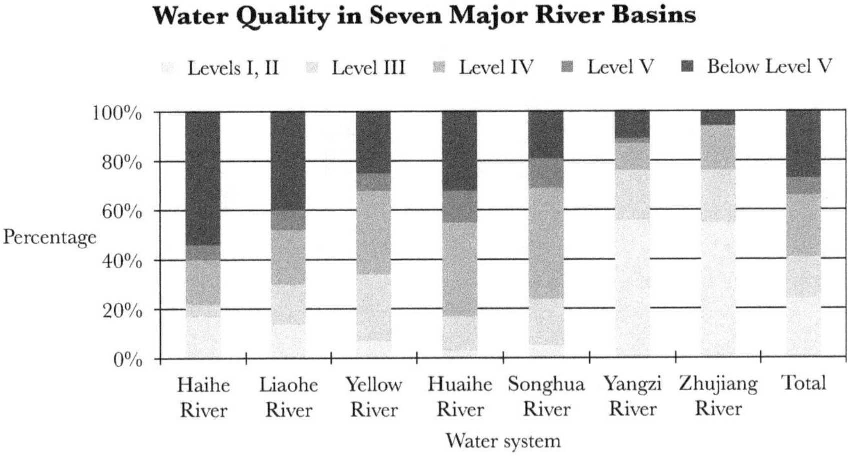

Figure A.7: Water quality in seven major river basins in 2005

Table A.7: Water quality in seven major river basins in 2005

Standards \ Water system	Haihe River	Liaohe River	Yellow River	Huaihe River	Songhua River	Yangzi River	Zhujiang River	Total
Levels I, II	17%	14%	7%	3%	5%	56%	55%	24%
Level III	5%	16%	27%	14%	19%	2%	21%	17%
Level IV	18%	22%	34%	38%	45%	11%	18%	25%
Level V	6%	8%	7%	13%	12%	2%	0	7%
Below Level V	54%	4%	25%	32%	09%	11%	6%	27%

Based on the results taken from 411 surface water monitoring stations based in the seven major river systems, quality attained the following ratings: Levels I–III 41%, Levels IV–V 32%, and Level V–27%. The Zhujiang and Yangtze River Basins enjoyed relative better water quality. On the other hand, the Huai River Basin is heavily polluted. The main pollution constituents are Ammonia Nitrogen (NH_3–N), Biochemical Oxygen Demand (BOD_5), permanganate,[3] and oil and petroleum derivatives.

[3] Potassium Permanganate and written as MnO_4. Among other uses, it can be added to water to treat and prevent various diseases afflicting fish.

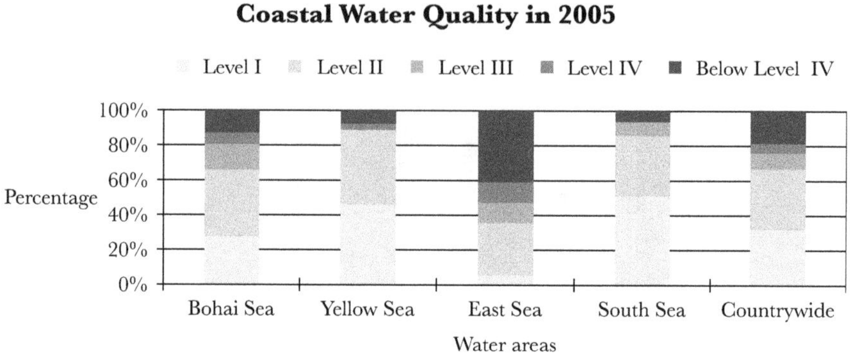

Figure A.8: Coastal water quality in 2005

Table A.8: Coastal water quality in 2005

Standards \ Water areas	Bohai Sea	Yellow Sea	East Sea	South Sea	Country-wide
Level I	27.70%	46.30%	5.40%	51.50%	32.10%
Level II	38.30%	42.60%	30.10%	34.30%	35.20%
Level III	14.90%	0	11.80%	8.10%	8.90%
Level IV	6.40%	3.70%	11.80%	0	5.50%
Below Level IV	12.80%	7.40%	40.90%	6.10%	18.40%

To improve these conditions, information on water quality should become more widely available and the public and advocate roles should expand.

Water quality in the Yellow Sea and South China Sea is relatively good; the quality of the Bohai Sea is average and quality of the East China Sea is rather poor. Coastal areas near Shanghai and Zhejiang Province are especially bad. The SEPA initiated the 'Clean Sea Water Campaign' in 2001, beginning with the outlets of the Yangtze and Pearl Rivers in 2005. The Campaign's aims to reduce nitrogen and phosphorus discharge from the Liaohe, Daling, Xiaoling, Haihe, Luanhe Rivers, as well as the provinces and cities along the Yellow River, by 15% from levels reached during the Tenth Five-year Plan period.

III. *Urban Areas*

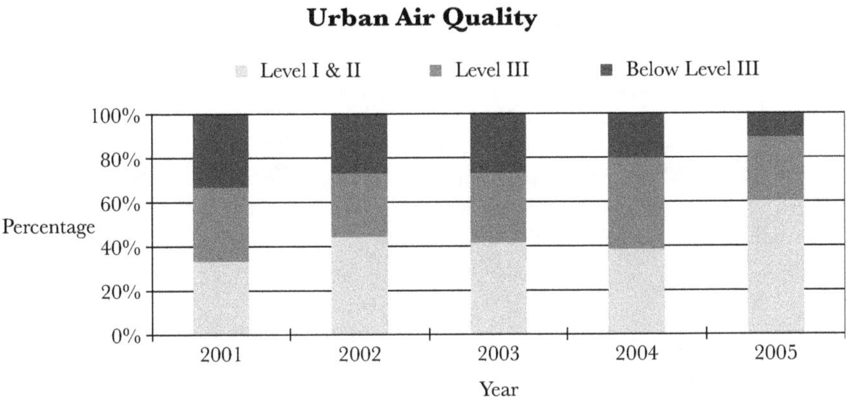

Figure A.9: Urban air quality in China during the Tenth Five-Year Plan period

Table A.9: Urban air quality in China during the Tenth Five-Year Plan period

Parameters \ Year	2001	2002	2003	2004	2005
Above Level II	33.40%	44.4%	41.7%	38.6%	60.3%
Level III	33.4%	28.7%	31.5%	41.2%	29.1%
Below Level III	33.2%	27.0%	26.8%	20.2%	10.6%

Air pollution takes a particularly heavy toll on urban dwellers. It also degrades farmland in when sulfur-tainted comes down as acid rain. Simply put, air pollution is one of the main environmental problems we face now.

Because numerous cities have not been monitored for many years, only a rough comparison of air quality conditions can be obtained. Using data from 2004, we can see that among 339 large cities, only 22.2% of these attained air quality at Level II. The number of cities with air quality Level III decreased by 24.1%; urban air quality has obviously improved. The main pollutant affecting urban air quality is PM 10, a class of particulate less than 10 micrometers in diameter; this can be easily inhaled by humans and can irritate the respiratory tract.

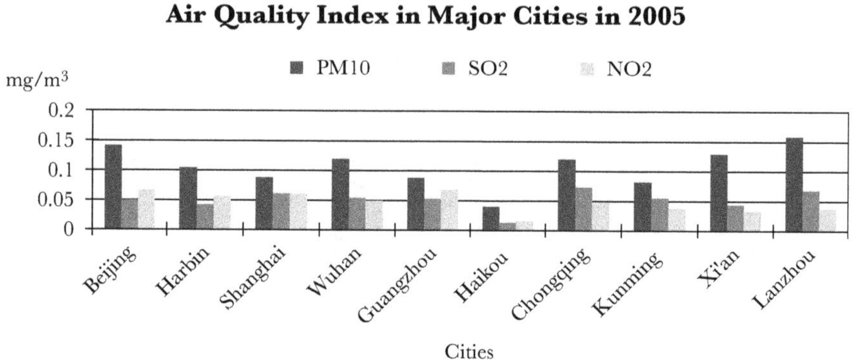

Figure A.10: Air quality in major Chinese cities in 2005

Table A.10: Air quality in major Chinese cities in 2005

Index \ Cities	Beijing	Harbin	Shanghai	Wuhan	Guangzhou	Haikou
PM10	14.1	10.4	8.8	11.9	8.8	4
SO_2	5	4.2	6.1	5.4	5.3	1.2
NO_2	6.6	5.6	6.1	5	6.8	1.5

Index \ Cities	Chong-qing	Kunming	Xi'an	Lanzhou
PM10	12	8.2	12.9	15.8
SO_2	7.3	5.5	4.4	6.8
NO_2	4.8	3.8	3.2	3.7

It has been implicated in the increase of asthma incidence in urban areas globally. Cities in Shanxi, Hebei, Gansu and Guizhou provinces experience the worst concentrations of PM 10.

Concentrations of SO_2 and NO_2 are relatively higher in more developed cities, since these gases largely originate from industrial and vehicular sources.

IV. *Solid Waste Processing and Disposal*

Industrial Solid Waste Generation and Disposal

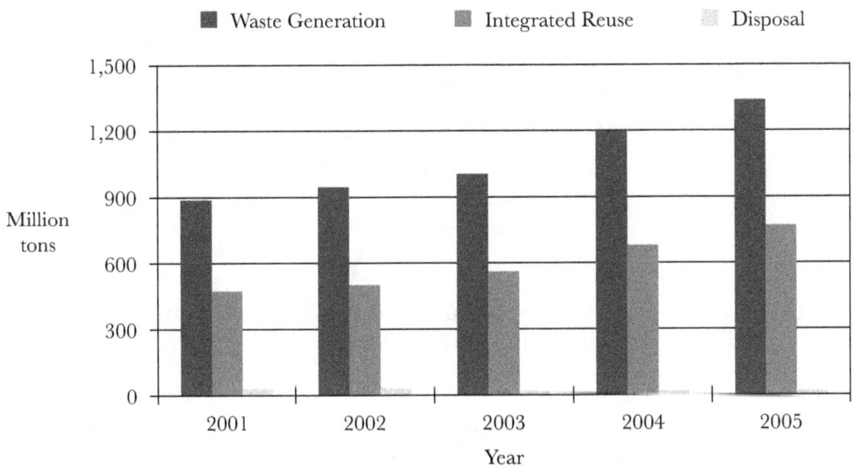

Figure A.11: Industrial solid waste generation and disposal in China during the Tenth Five-Year Plan period

Table A.11: Industrial solid waste generation and disposal in China during the Tenth Five-Year Plan period (1,000,000 tons)

Year / Disposal Intensity	2001	2002	2003	2004	2005
Generation	887	945	1004.28	1200	1340
Integrated Reuse	473	500.6	560.4	677.95	770
Disposal	28.93	26.35	19.4	17.92	16.54

V. *Economic Development and Energy Consumption*

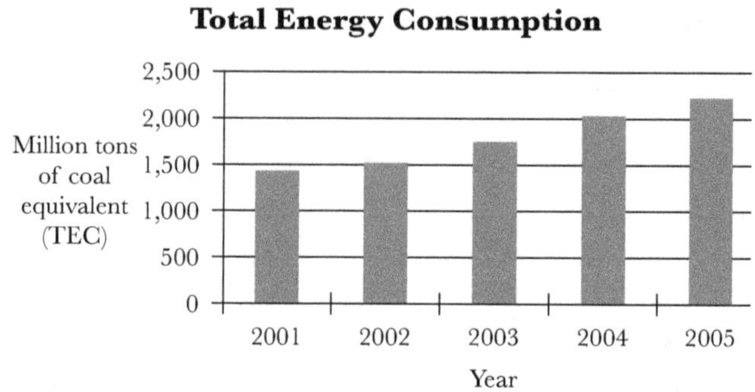

Figure A.12: Total energy consumption in China during the Tenth Five-Year Plan period (1,000,000 tons coal equivalent)

Table A.12: Total energy consumption in China during the Tenth Five-Year Plan period (1,000,000 tons coal equivalent)

Region \ Year	2001	2002	2003	2004	2005
Eastern China (10^6 TEC)	1,430	1,517	1,749	2,032	2,224

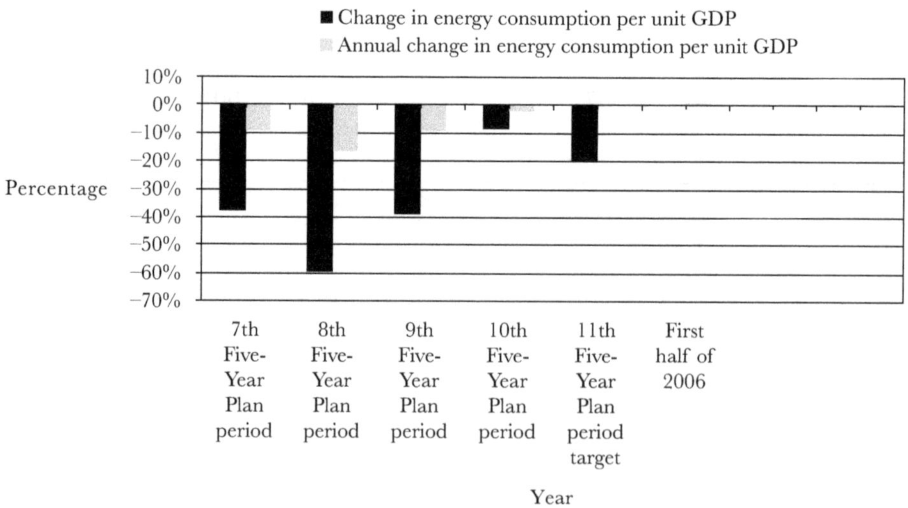

Figure A.13: Decrease in energy consumption per unit of GDP after the Seventh Five-Year Plan period

Table A.13: Decrease in energy consumption per unit of GDP after the Seventh Five-Year Plan period[4]

Percentage \ Time	The Seventh Five-Year Plan period	The Eighth Five-Year Plan period	The Ninth Five-Year Plan period	The Tenth Five-Year Plan period	The goal of the Eleventh Five-Year Plan period	The first half of 2006
The change in energy consumption per unit GDP	–38%	–59.50%	–39%	–8.50%	–20%	–
The annual change in energy consumption per unit GDP	–8.80%	–16.20%	–9%	–1.80%	–	0.80%

The Eleventh Five-Year Plan set the target that energy consumption per unit of GDP in 2010 would be 20% than the ratio at the end of 2005. To meet this goal, energy consumption per unit of GDP ought to decrease 4.4% annually from 2006–2010. Can this goal be achieved? Can the high-input, high-energy use, low-efficiency economic model currently at work in China begin to change in this period?

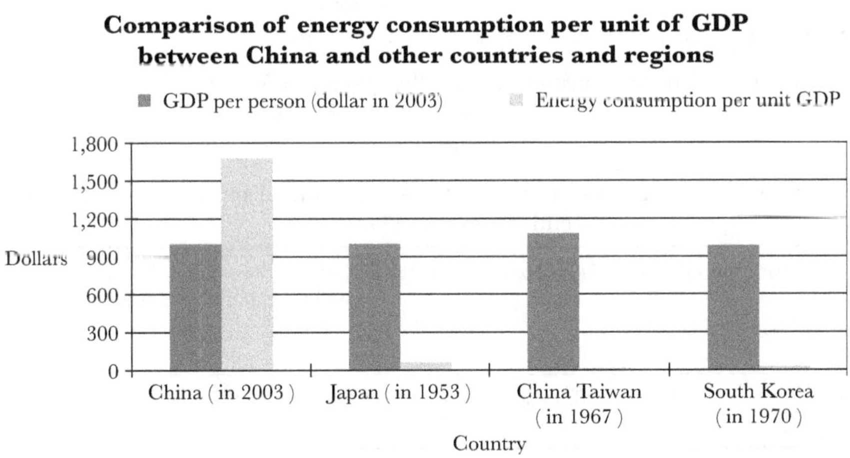

(Source: The Institute of Energy Economics, Japan and SYWG Research & Consulting)

Figure A.14: Comparison of energy consumption per unit of GDP between China and other countries and regions

[4] Note that in the first half of 2006, this metric outpaced economic growth.

Table A.14: Comparison of energy consumption per unit of GDP between China and other countries and regions (USD)

Country / GDP	China 2003	Japan 1953	China Taiwan 1967	South Korea 1970
GDP per person	1000	1000	1080	985
Energy consumption per unit GDP	1680	60	10	25

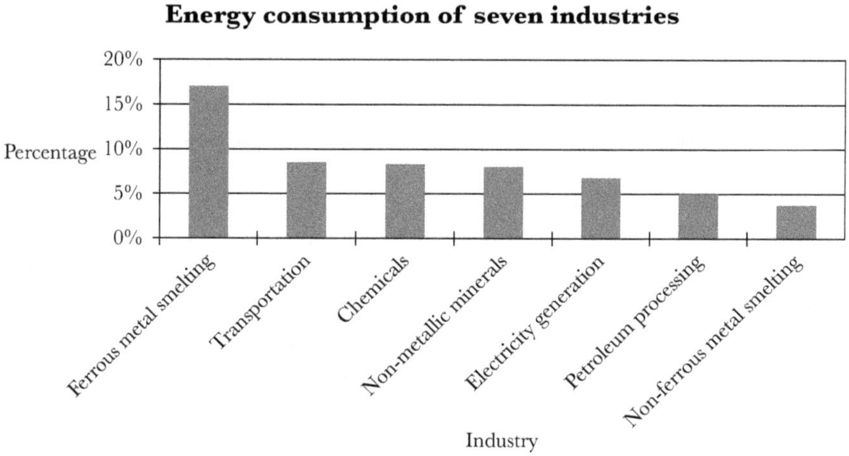

(Source: World Bank and SYWG Research & Consulting)

Figure A.15: Energy consumption of the seven major industries during the Tenth Five-Year Plan period

Table A.15: Energy consumption of the seven major industries during the Tenth Five-Year Plan period

Industry / Energy consumption	Ferrous metal smelting	Transportation	Chemicals	Non-metallic minerals	Electricity generation	Petroleum processing	Non-ferrous metal smelting
Energy consumption	17%	8.50%	8.30%	8%	6.80%	5%	3.70%

(Source: World Bank and SYWG Research & Consulting)

During the Tenth Five-Year Plan the extensive economic growth model continued to move along unchanged.[5] Since the end of 2002, energy and input-intensive industries, such as electricity generation, metallurgical manufacturing, and construction have grown at breakneck speed, in some cases attaining annual growth rates of 15%. This rate of growth and method of production places a heavy burden on the environment.

VI. *Financing Green Initiatives*

The first Chinese Green GDP Statistical Public Report published in September 2006 made a significant break with standard economic accounting systems by including resource and environmental costs in the national economical accounting system. Among other eye-opening data, the report delivered the following figures on various costs attributed to pollution: air pollution, 219.8 billion RMB; water and solid waste pollution, 286.2 billion RMB; and pollution accidents, 5.74 billion RMB. These categories make up 43%, 56% and 1% of the total costs.

Even as China's continued high economic growth rates stole the headlines, environmental degradation continues to be a less widely reported, but still important, story. The problems unearthed in the GDP provide a more balanced view of standard economy economic growth and environmental conditions. This new perspective promises to be a hot topic of discussion among both Chinese and foreigners this year.

---

[5] The "extensive" growth model is distinct from the 'intensive' growth model in that the former emphasizes a high-input, high volume, low-margin, low-efficiency production system; the intensive model reverses these conditions, primarily by moving into higher-margin, knowledge-economy forms of production and distribution.

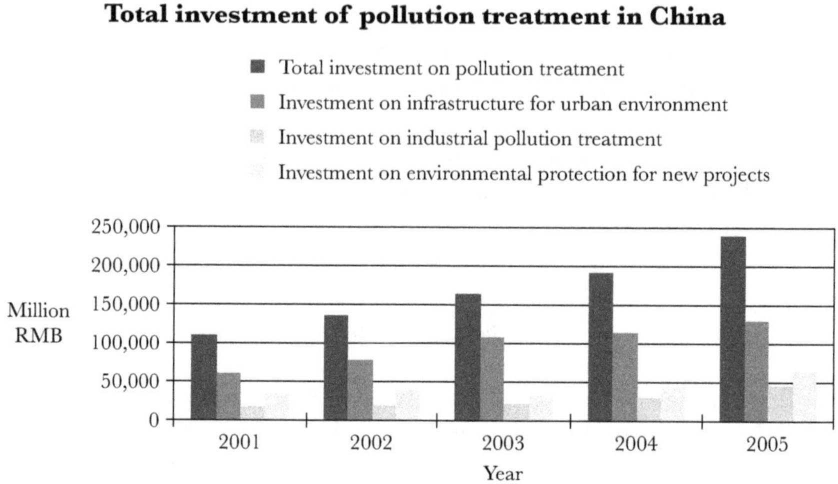

Figure A.16: Total investment of pollution treatment in China during the Tenth Five-Year Plan period

Table A.16: Total investment of pollution treatment in China during the Tenth Five-Year Plan period

Investment \ Year	2001	2002	2003	2004	2005
Total investment on pollution treatment	1106.6	1363.4	1627.3	1908.6	238.8
Investment on infrastructure for the urban environment	595.7	785.3	107.2	114.0	1289.7
Investment on industrial pollution treatment	174.5	188.4	221.8	308.1	458.2
Investment on environmental protection for new projects	336.4	389.7	333.5	460.5	640.1

Percentage of GDP invested in pollution treatment annually

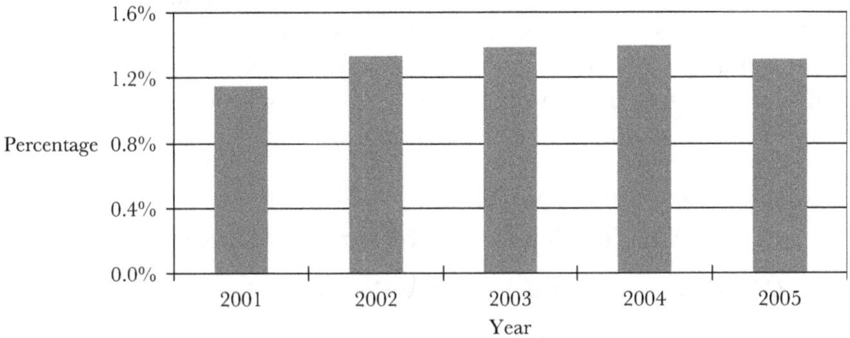

Figure A.17: Percentage of GDP invested in pollution treatment annually during the Tenth Five-Year Plan period

Table A.17: Percentage of GDP invested in pollution treatment annually during the Tenth Five-Year Plan period in the developed East

	2001	2002	2003	2004	2005
Eastern China	1.15%	1.33%	1.39%	1.40%	1.31%

Investments in environmental initiatives have increased significantly in recent years. The percentage of GDP invested in pollution treatment 2001 measured 1.15%, but jumped to 1.33% in 2002, essentially remaining unchanged over the remaining years of the Tenth Five-year Plan: 2003 (1.39%), 2004 (1.40%) and 2005 (1.31%). The structure of fiscal policy in this regard changed as well, with multiple revenue sources come to bear on environmental allocations. From 2001–2005, expenditures reached 111.9 RMB, with 108.3 billion RMB coming from the sale of bonds to finance the national debt. Several individual initiatives absorbed the lion's share of these funds: sandstorm prevention and treatment near Beijing and Tianjin; natural forest protection; administration of the farmland-to-forest program;[6] water pollution treatment in the Three Gorge Gorges area, including the riparian

[6] This program also includes a related component that returns farmland to grassland.

environment upriver from the dam's reservoir; operation of the 'Three Rivers and Three Lakes' pollution abatement program; and wastewater treatment and reuse alongside solid waste disposal.

2006 marked another milestone: including of environmental protection allocations in the national budget. Changes in fee structures and taxes also had an impact on the national budget.

VII. *The Greenhouse Effect and CO_2 Discharge*

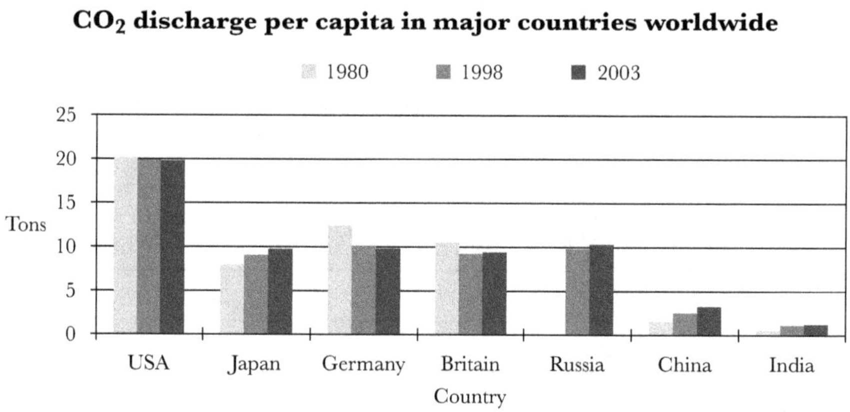

Figure A.18: CO_2 discharge per capita in major countries worldwide

Table A.18: CO_2 discharge per capita in major countries worldwide (tons)

Year \ Country	USA	Japan	Germany	Britain	Russia	China	India
1980	20.1	7.9	12.4	10.5	–	1.5	0.5
1998	19.9	9	10.1	9.2	9.8	2.5	1.1
2003	19.8	9.7	9.8	9.4	10.3	3.2	1.2

(Source: UNDP Human Development Report 2006)

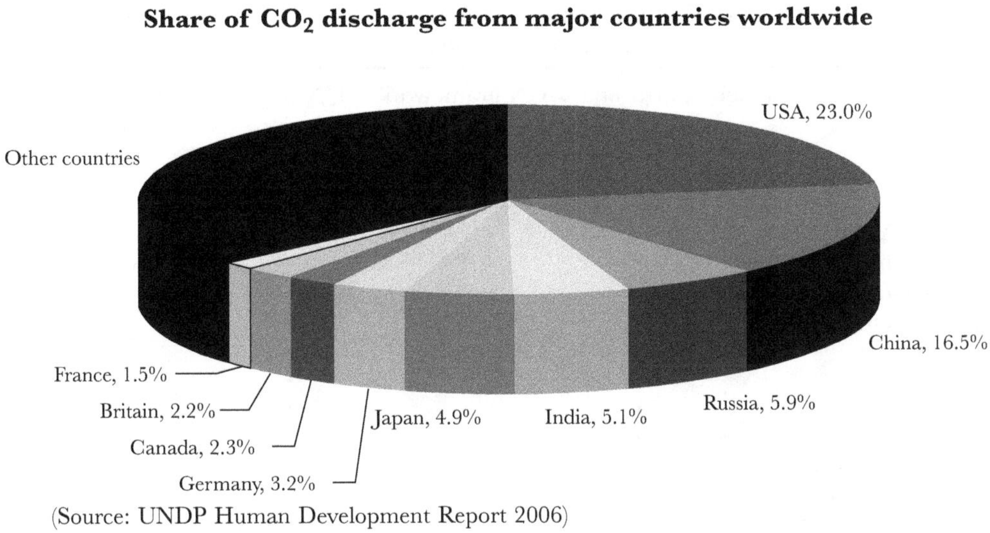

Figure A.19: Share of CO_2 discharge from major countries worldwide

CO_2 emissions are a leading contributor to global warming. China ranks number two in total CO_2 emissions behind the United States, but its per capita output falls far short of figures for developed countries like the U.S.A., Japan, and the European Union nations. Moreover, China does not exceed its targets specified in the Kyoto Protocol. However, with new restrictions due as a result of international negotiations, China will faced increasing pressure to reduce its greenhouse gas output. Clearly, China still has a great deal of work to do to meet its responsibilities.

VIII. *Miscellaneous*

On April 27, 2005, the Chinese State Council ratified the Cartagena Protocol on Biosafety, and filed the ratified document with the UN Headquarters on June 8, 2005. The Protocol went into effect in China on September 6, 2005.

Table A.19: Status of international environmental laws by key nations worldwide (as of December 2006)

Country \ Protocol	Cartagena Protocol on Biosafety	UN Framework Convention on Climate Change (UNFCCC)	UNFCCC Kyoto Protocol	Convention on Biodiversity
USA	–	•	◦	◦
China	•	•	•	•
Japan	•	•	•	•
Britain	•	•	•	•
France	•	•	•	•
Russia	–	•	•	•
Italy	•	•	•	•
India	•	•	•	•
South Korea	–	•	•	•
Germany	•	•	•	•

◦ signed
• approved, adopted, ratified, joined

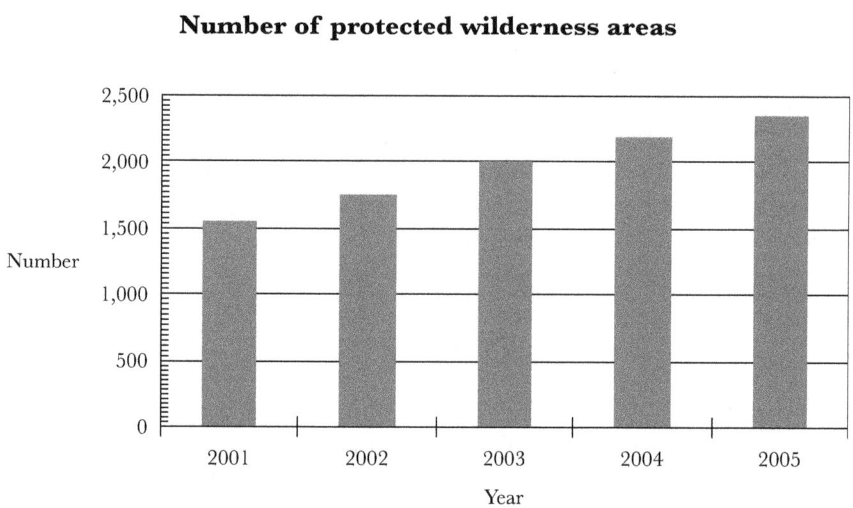

Figure A.20: Number of protected wilderness areas in China during the Tenth Five-Year Plan period

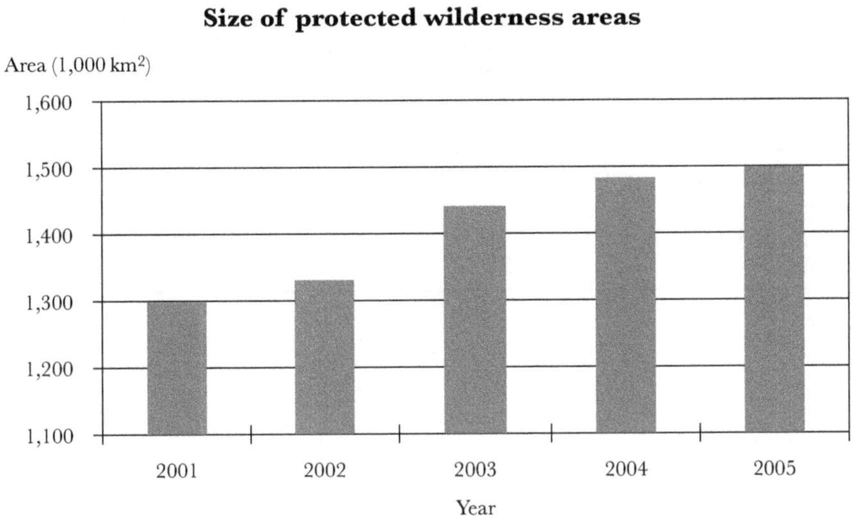

Figure A.21: Size of protected wilderness areas in China during the Tenth Five-Year Plan period

The Chinese government regards organizing nature preserves as one of its most important metrics of ecological protection. Through the end of 2005, there were 2,349 protected areas in China covering a total area of 1.5 million km^2, or about 15% of China's total land area. In other words, a network of protected wilderness areas covering the full range of habitat types found in China has been established. Consequently, 85% of land ecosystems types, 85% of the available wild fauna communities, and 65% of natural flora communities have been put under protection.[7]

[7] Note the qualifiers; this is not to say that 85% of the land has been brought under protection, but rather, 85% of the various ecological system types have at least one such area protected. The same is true for fauna and flora species. While much of the population remains unprotected, 85% and 65% of species have some part of their populations under protection.

Table A.20: Number of protected wilderness areas in China during the Tenth Five-Year Plan period (1,000 km^2)

Indexes \ Year	2001	2002	2003	2004	2005
The number of protected areas	1551	1757	1999	2194	2349
The size of protected areas	129.89	132.95	143.98	148.226	149.949

INDEX

www.ingramcontent.com/pod-product-compliance
Lightning Source LLC
LaVergne TN
LVHW081514300926
849161LV00006B/1170
* 9 7 8 9 0 0 4 1 6 8 0 0 8 *